From MODERNITY TO COSMODERNITY

SUNY series in Western Esoteric Traditions

David Appelbaum, editor

From MODERNITY to COSMODERNITY

Science, Culture, and Spirituality

BASARAB NICOLESCU

STATE UNIVERSITY OF NEW YORK PRESS

Published by
STATE UNIVERSITY OF NEW YORK PRESS, ALBANY

© 2014 State University of New York

All rights reserved

Printed in the United States of America

No part of this book may be used or reproduced in any manner whatsoever without written permission. No part of this book may be stored in a retrieval system or transmitted in any form or by any means including electronic, electrostatic, magnetic tape, mechanical, photocopying, recording, or otherwise without the prior permission in writing of the publisher.

For information, contact
State University of New York Press, Albany, NY
www.sunypress.edu

Production, Laurie D. Searl
Marketing, Fran Keneston

Library of Congress Cataloging-in-Publication Data

Nicolescu, Basarab.
  From modernity to cosmodernity : science, culture, and spirituality / Basarab Nicolescu.
    pages cm. — (SUNY series in western esoteric traditions)
  Includes bibliographical references and index.
  ISBN 978-1-4384-4963-0 (hc : alk. paper) 978-1-4384-4964-7 (pb : alk. paper)
  1. Reality. 2. Complexity (Philosophy) 3. Religion and science. I. Title.

BD331.N488 2014
110—dc23   2013006675

10 9 8 7 6 5 4 3 2 1

CONTENTS

LIST OF ILLUSTRATIONS	ix
INTRODUCTION	1
CHAPTER ONE	
FROM SHATTERED CULTURE TOWARD TRANSCULTURE	3
The Christian Origin of Modern Science	3
Do Science and Culture Have Something in Common?	7
The Transcultural and the Mirror of the Other	10
The Transreligious Attitude and the Presence of the Sacred	14
CHAPTER TWO	
CONTEMPORARY PHYSICS AND THE WESTERN TRADITION	19
Tradition and Traditions	19
Science and Tradition: Two Poles of a Contradiction	20
A Possible Bridge between Sciences and Tradition:	
The Rationality of the World	23
Describing God's Being. . . .	24
Movement and Discontinuity: The Eternal Genesis of Reality	26
Scientific Thinking and Symbolic Thinking: Icons and Thêmata	30
A Necessary Encounter	34
CHAPTER THREE	
THE GRANDEUR AND DECADENCE OF SCIENTISM	37
The Classical Vision of the World and the Death of Man	37
Modern Mahabharata-like Drama: The Quantum Vision of the	
World	40
CHAPTER FOUR	
THE VALLEY OF ASTONISHMENT: THE QUANTUM WORLD	45
About the Difficulties of the Journey	45
Planck, Discontinuity, and the Quantum Revolution	47
The Particle and Quantum Spontaneity	50

Heisenberg's Relations and the Failure of Classic Determinism 52
The Multiplicity of Quantum Values and the
  Role of Observation 53
Quantum Vacuum: A Full Vacuum 54
Quantum Nonseparability 56

CHAPTER FIVE
THE ENDLESS ROUTE OF THE UNIFICATION OF
THE WORLD 59

Is a Single Energy the Source of the World's Diversity? 59
The Final Theory: Superstrings? 62
The Unification of Heaven and Earth 63
Can Everything Be Unified? 65
Everything Is Vibration 66
The Mystery Theorists 70
Seekers of Truth 71

CHAPTER SIX
THE STRANGE FOURTH DIMENSION 75

CHAPTER SEVEN
THE BOOTSTRAP PRINCIPLE AND THE UNIQUENESS
OF OUR WORLD 87
Eddington and the Epistemological Principles 87
Unity and Self-Consistency: The Bootstrap Principle 88
Is There a Nuclear Democracy? 91
The Bootstrap and the Anthropic Principle 92
Methodological Considerations 95

CHAPTER EIGHT
COMPLEXITY AND REALITY 99
The Emergence of Complex Plurality 99
Some Reflections on Systemic Thinking 101
Systemic Thinking and Quantum Physics 102
Levels of Reality 104
Is There a Cosmic Bootstrap? 107
Evolution and Involution 110

CHAPTER NINE
THE HUMAN BEING: THE MOST PERFECT OF ALL SIGNS 113
Natural Language and Scientific Language 113
Peirce and Spontaneity 114

Invariance and Thirdness 116
The Possibility of a Universal Language 118

CHAPTER TEN
BEYOND DUALISM 121

A Stick Always Has Two Ends 121
Stéphane Lupasco (1900–1988): The Herald of the Coming Third 125
The Included Third 127
The Ternary Dialectics of Reality 128
Triadic Systemogenesis and the Three Matters 130
Nonseparability and the Unity of the World 131
The Nature of Space-Time 131
Is Lupasco a Prophet of the Irrational? 132
The Experienced Third 134

CHAPTER ELEVEN
THE PSYCHOPHYSICAL PROBLEM 137

Reduction and Reductionism 137
The *Coincidentia Oppositorum* and Hermetic Irrationalism 138
The Core of the Problem: We Are Too Deeply Immersed
  in the Seventeenth Century 139
The Most Important Task of Our Time: A New Idea about Reality 142
New Perspectives in the Ternary-Quaternary Debate 142
Umberto Eco's Logical and Epistemological Error 144

CHAPTER TWELVE
FROM THE QUANTUM WORLD TO IONESCO'S ANTITHEATER
AND QUANTUM AESTHETICS 147

For a Yes or for a No 147
Ionesco and the Non-Aristotelian Theater 148
Gregorio Morales: Quantum Aesthetics and Quantum Theater 151

CHAPTER THIRTEEN
THE THEATER OF PETER BROOK AS A FIELD OF
STUDY OF ENERGY, MOVEMENT, AND INTERRELATIONS 155

CHAPTER FOURTEEN
FROM CONTEMPORARY SCIENCE TO THE WORLD OF ART 167

André Breton and the Logic of Contradiction 167
Georges Mathieu and Aristotle's Cage 169
Salvador Dali and Nuclear Mysticism 170
Frédéric Benrath, Karel Appel, and René Huyghe 174

CHAPTER FIFTEEN
VISION OF REALITY AND REALITY OF VISION — 177
- Poincaré and Sudden Enlightenment — 178
- Hadamard and Thinking without Words — 180
- Kepler and the Living Earth — 182
- Bohr and Complementarity — 184
- Understanding the Reality of the Imaginary: The Imaginary and the Imaginal — 186

CHAPTER SIXTEEN
CAN SCIENCE BE A RELIGION? — 189
- The Clowns of the Impossible — 189
- Highlights of the New Barbarity — 190
- Between the Anecdote and the Unspeakable — 191
- The Sokal Affair: Beyond Three Extremisms — 193
- A Necessary Isomorphism — 197
- The End of Science? — 198
- The Spiritual Dimension of Democracy: Utopia or Necessity? — 199

CHAPTER SEVENTEEN
THE HIDDEN THIRD AND THE MULTIPLE SPLENDOR OF BEING — 203
- Premodernity, Modernity, Postmodernity, and Cosmodernity as Different Visions of the Relation between the Subject and the Object — 203
- Ladder of Divine Ascent and Levels of Being — 205
- Toward a Unified Theory of Levels of Reality — 207
- At the Threshold of New Renaissance — 214

NOTES — 217

BIBLIOGRAPHY — 239

NAME INDEX — 253

SUBJECT INDEX — 259

LIST OF ILLUSTRATIONS

Figure 17.1	The Relationship between Subject and Object in Premodernity	204
Figure 17.2	The Relationship between Subject and Object in Modernity	204
Figure 17.3	The Relationship between Subject and Object in Postmodernity	204
Figure 17.4	The Relationship between Subject and Object in Cosmodernity	210
Figure 17.5	Trans-Reality	211

INTRODUCTION

The word *reality* is one of the most ambiguous words in all languages. We all think that we know what reality is, but if we were asked, we would find that there are as many meanings of this word as there are people on earth. Therefore, it is not surprising that so many conflicts continuously disturb individuals and nations: the reality of one against the reality of the other.

Under these conditions, it is almost miraculous that the human species still exists. The explanation is relatively simple: a statistical kind of faith in what reality is at some moment is created as an effect of technoscience. Thus, the dominant concept of reality during the last century was based on classical science. It used to reinforce our belief that we were living in a rational, deterministic, and mechanistic world, destined to an endless progress. The unexpected event of September 11, 2001, shattered this faith of modernity.

However, the triple revolution that spanned the twentieth century—the quantum revolution, the biological revolution, and the information revolution—should have thoroughly changed our view of reality. And yet, our mentalities remained unchanged. The massacre of humans by humans increased endlessly. The old viewpoint remains dominant. What is the cause of this blindness?

In this book, I assume the statement made in 1948 by Wolfgang Pauli, Nobel laureate in physics and one of the founders of quantum mechanics: "the formulation of a new idea of reality is the most important and most difficult task of our time." More than sixty years later, this task remains unfulfilled. I hope that the present book makes a step forward toward the accomplishment of this task.

The contemporary growth of knowledge is without precedent in human history. We have explored otherwise unimaginable levels: from the infinitely small to the infinitely large, from the infinitely brief to the infinitely long. The sum of knowledge about the universe and natural systems accumulated during the twentieth century far surpasses all that was known during all other centuries combined. How is it that we know more about what we do and less about who we are? How is it that the accelerating proliferation of disciplines makes the unity of knowledge more and more impossible to even imagine? How is it that as the exterior universe becomes more known, the meaning of our life and of our death declines into insignificance, even absurdity? Must atrophy of interior being be the price we pay for techno-scientific advance? Must the individual and social happiness that scientism first promised us recede indefinitely, like a mirage?

To be sure, the potential triple self-destruction—material, biological, and spiritual—is the product of a blind but triumphant technoscience, obedient only to the implacable logic of utilitarianism. But how can we expect the blind to see?

Paradoxically, everything is in place for our self-destruction, but everything is also in place for positive change, just as there has been at other great turning points in history. The challenge of self-destruction has its counterpart in the hope of self-birth. The global challenge of death has its counterpart in a visionary, transpersonal, and planetary consciousness, which could be nourished by the miraculous growth of knowledge.

Of course, my practice of more than forty years in theoretical elementary particle physics is crucial in the writing of the present book. It is this experience and knowledge that led me to question the meaning of science in its interaction with spirituality. Also, the deep contact with my PhD students in transdisciplinarity at the University Babeș-Bolyai from Cluj-Napoca in Romania stimulated my thinking a great deal.

I am very grateful to the Stellenbosch Institute of Advanced Study (South Africa) for a fellowship, which allowed me to dedicate all of my time to writing the present book.

I also thank Karen-Claire Voss and Andy Green for their invaluable help in correcting different versions of the manuscript and for suggestions in the clarification of some ideas. The contribution of Anamaria Brailean in the preparation of the first form of the manuscript is acknowledged.

I am most grateful to the great Spanish poetess Clara Janés for the authorization of publishing her beautiful poem *The Hidden Third* in the present book.

I dedicate this book to the memory of my dear friend and collaborator Paul Cilliers (1957–2011), professor of philosophy at Stellenbosch University and a major world figure as theorist and philosopher of complexity, who died suddenly on July 31, 2011. I am sure that he will be delighted to see this book in print, wherever he is now.

CHAPTER ONE

FROM SHATTERED CULTURE TOWARD TRANSCULTURE

THE CHRISTIAN ORIGIN OF MODERN SCIENCE

Modern science, and even modernity itself, began in Europe in 1632, with the publication of Galileo's book *Dialogue Concerning the Two Chief World Systems*.[1]

The identity of modern science is, of course, crucial for Europe. Its past identity has many embodiments: the Greek, Roman, and Christian legacies; the Reformation; the Renaissance and the Age of Enlightenment; the modern messianisms (communism and nazism); and psychoanalysis. This contradictory relationship between the unity and diversity of its identity is both an asset and a challenge.

There is one often-neglected aspect that nevertheless seems important when trying to decipher the future identity of Europe: the birth of modern science there, in Europe, at the time of its great founders—Galileo, Newton, and Kepler. Of course, other civilizations—Greek, Egyptian, Sumerian, Mayan, Arabian, and Chinese—made crucial contributions to science in its general meaning. For example, the Chinese civilization had everything necessary to give birth to modern science—theoretical thinking and technology—but the historical fact is that it did not do so. The point is that the Chinese vision of the cosmos forbade them to accept fragmentation and separation.

Joseph Needham asked in his famous book *Chinese Science and the Occident*: "why is modern science, as a mathematization of the hypotheses regarding nature, with all its implications in the field of advanced technology, only rapidly increasing in the West, in the era of Galileo?"[2]

The search for the most intimate resorts, of the thinking style and of the imaginary that leads to a certain worldview typical of a certain era, is indispensable for a rigorous approach to Joseph Needham's question. Europe's cultural and spiritual environment contained the germ of modern science.

This is the premise that led me in 1988 to formulate the hypothesis of *the Christian origin of modern science* in my book *Science, Meaning, & Evolution: The Cosmology of Jacob Boehme*.[3]

According to my analysis, the dogmas of the Christian Trinity and the Incarnation, which permeated the cultural and spiritual environment of Galileo's time, accompanied by an important technological development, enabled the formulation of the methodology of modern science. The fact that Christianity was the dominating religion in Europe was also an important historical aspect.

After the publication of my book, I discovered a short study by Alexandre Kojève (1902–1968) entitled "L'origine chrétienne de la science moderne."[4] Kojève, a known neo-Marxist French philosopher of Russian origin, cannot be remotely suspected of sympathy toward Christianity. The main argument of his study, which was published in 1964, is based uniquely on the dogma of the Incarnation.

Kojève starts by acknowledging the birth of mathematical physics in Western Europe in the sixteenth century. For Kojève, pagan theology represents the theory of the double transcendence of God: "The *Theos* of 'classic' paganism . . . is not only beyond the world where the pagan lives. This *Theos* is irremediably beyond the *Beyond* that the pagan may, at some point, access after his death."[5] Double transcendence is the conception of an unsurpassable wall that does not allow us to conceive perfection in our world, perfection that is, nevertheless, assumed by the laws of physics. According to Kojève, this perfection is ensured by mathematical physics, "Therefore, for pagans like Plato and Aristotle, as for all civilized Greeks who are therefore likely to pursue the sciences, looking for a science like modern *mathematical physics* would be not only madness, but a huge scandal—as, indeed, for the Jewish people."[6] As for the obvious reproach regarding the conflict between the founders of modern science and the Catholic Church, Kojève responded with justification: "What these scientists are fighting against is scholasticism in its most advanced stage; that is, the Aristotelianism portrayed in its pagan authenticity, whose incompatibility with Christian theology had been clearly observed and indicated by the first precursors of the philosophy of modern times."[7] Furthermore, Kojève reviewed the Christian dogmas of God's uniqueness, of *ex nihilo* creation, of the Trinity and the Incarnation, in order to conclude for the preeminence of the latter dogma, that of the Incarnation, in the birth of modern science. He eliminated the first two dogmas simply because they were also found elsewhere, namely in Judaism and Islamism. He also removed the dogma of the Trinity because, wrote Kojève, "it rather incites 'mystical' introspection or metaphysical speculations than a careful observation of a body of phenomena and of their experimental manipulations."[8] What remained was the dogma of the Incarnation, about which Kojève wrote: "Indeed, what is the Incarnation, if not the opportunity for God to be effectively present in the temporal world, in which we live ourselves, yet without decaying from his absolute perfection?"[9] For Kojève, Copernicus "lifted to Heaven the body of Christ, revived by the entire terrestrial world where Jesus died, after having been born here. But whatever this heaven might have been for the faithful Christians, it was but a 'mathematical' or a mathematizable sky for all the savants of the era."[10]

Kojève's theses could seduce an important medieval literature specialist such as Alexandre Leupin. In his book *Fiction et Incarnation*,[11] published in 1993, Leupin showed that the pagan-Christian epistemological rupture introduced a new conception about language and a new regime of truth and fiction. Christian epistemology requires a realism of language and concepts. Language structure is not caught in an infinite loop, closed on itself. It opens toward reality, toward meaning. It refers to an event of reality. Leupin operates seductive conjunctions: concept/divine truth, experience/life of Christ, event/writing. The contradictory pair God-out-of-the-world/God-from-this-world is foundational for modern science, which is characterized by another contradictory pair: mathematics/scientific experiment. Leupin says: "Biblical writing is co-substantially the story of the thing (*res gestae*) and the thing itself, the unthinkable identity, which places the entire medieval writing, whatever its nature, under the absolute sign of the impossible. At the same time holding the experience and being the experience itself, the truth and the representation of the truth, the testamentary 'anti-writing' promotes writing to a dignity that is unrelated to the values that had been attributed to it by pagan antiquity."[12]

Of course, Kojève's theses were violently criticized by the American philosopher Steven Louis Goldman,[13] who accused him of having assigned the paternity for the mathematical physics to Christianity.

Kojève is, in fact, right and wrong at the same time.

He was certainly right when he stressed the importance of the dogma of the Incarnation for the birth of modern science.

But he was wrong in first reducing modern science to mathematical physics.

Modern science is defined by its *methodology*, which was formulated by Galileo in the form of three postulates that are still valid today (see chapter 3):

1. The existence of universal laws of mathematical nature.

2. The discovery of these laws through scientific experimentation.

3. The perfect reproducibility of experimental results.

A careful examination of these postulates shows that mathematics certainly plays an important role, although not an exclusive one. Mathematics is at the same time an artificial language, different from familiar language, but also, according to Galileo, expressed by Salvicio, a common language between God and humans.[14] If the first postulate and, perhaps, the third have some relation with the dogma of the Incarnation, this is not obvious for the second postulate, which introduces a *third* term in the human being–God relationship: nature.

Kojève was also wrong in having neglected the role of the dogma of the Trinity in the birth of modern science. It is certainly wrong to reduce this dogma to an "incitement towards mystical introspection" and to "metaphysical speculations." As shown in a study conducted in collaboration with Catholic priest Thierry Magnin, philosopher and poet Michel Camus, and historian of religion Karen-Claire Voss,[15] the dogma of the Trinity operates with a different logic from that of classical logic—the logic of the included middle, studied by Stéphane Lupasco (1900–1988).[16] This logic is an extraordinary tool for analysis and inference and is necessary for pondering the harmonious coexistence of the opposites. It is not accidental that the logic of the included middle is the one that solves all the paradoxes of quantum mechanics and quantum physics as well as the one that can give a rational explanation of Christian texts.[17]

In this context, the example of Jacob Boehme (1575–1624), whom Hegel called "the first German philosopher" and whose work has exerted an undeniable influence on Newton, Novalis, Schlegel, Goethe, Fichte, Schelling, and Karl Marx, is very significant.

In order to explain the world, Boehme invented a septenary typology inspired directly by the dogmas of the Incarnation and the Trinity. The dogma of the Trinity consists of three terms that in turn have a ternary structure. Thus, Boehme obtains a structure with nine elements, two of which are discontinuities (*Fiat*, in Bohme's language). If the ternary (associated with the dogma of the Trinity) concerns the inner dynamics of each system, the septenary (associated with the dogma of the Incarnation) is the foundation of the manifestation of all processes.[18]

The hypothesis of the Christian origin of modern science opens an interesting path to a new vision of what the roots and the future of Europe are. If Christianity is really the origin of modern science, it is very hard to believe that science has nothing in common with culture and spirituality. It is like saying that a baby has nothing in common with his or her mother. The fact that many scientists are atheist or agnostic is irrelevant in this context. Science itself is one thing; scientists are another thing. A given scientist is the conscious side of science and the historical movement and evolution of science are its subconscious part, like two sides of the same coin.

DO SCIENCE AND CULTURE HAVE SOMETHING IN COMMON?

At the beginning of human history, science and culture were inseparable. They were animated by the same questions, those about the meaning of the universe and the meaning of life.

In the Renaissance, those ties were not yet broken. As its name indicates, the first university was dedicated to studying the universal. The universal was embodied in those who would make their stamp on the history of knowledge. Gerolamo Cardano (1501–1576), the inventor of imaginary numbers and of the suspension system that bears his name, was a mathematician, a doctor, and an astrologer: the same person who established the horoscope of Christ was the author of the first systematic exposition of the calculus of probabilities.[19] Johannes Kepler (1571–1630) was both an astronomer and an astrologer.[20] Isaac Newton (1643–1727) was simultaneously a physicist, a theologian, and an alchemist. He was as captivated by the Trinity as by geometry, and he spent more time in his alchemical laboratory than in the elaboration of his *Philosophiae Naturalis Principia Mathematica.*[21] The founders of modern science had nothing in common with the stereotypical image of a scientist. Paradoxically, the scientist is forced, in spite of himself, to become a high priest of truth, an embodiment of rigor and objectivity. The complexity of the birth of modern science and of modernity itself helps us to understand the subsequent complexity of our own time.

The germ of the split between science and meaning, between subject and object, was certainly present in the seventeenth century, when the methodology of modern science was formulated, but it did not become full-blown until the nineteenth century, when the disciplinary big bang took flight.

In our time, the split was consummated. Science and culture have nothing more in common. This is why one speaks of science *and* culture. Every self-respecting government has a minister of culture and a minister of science. Every self-respecting international higher educational institution has a Department of Culture and a Department of Sciences. Those who try to cross the frontiers discover the risks of such an adventure. Science does not have access to the nobility of culture, and culture does not have access to the prestige of science.

In Europe, within science, one distinguishes the exact sciences from the human sciences, as if the exact sciences were inhuman (or subhuman) and the human sciences inexact (or non-exact). Anglo-Saxon terminology is still worse: one speaks of hard sciences and soft sciences. We will pass over the sexual connotation of these terms in order to explore their meaning. What are at stake are the ideas of definition, rigor, and objectivity, which convey the sense of exactitude (or of "hardness"). According to classical thought, the only exact definition is a mathematical definition, the only rigor worthy of its name is mathematical rigor, and the only objectivity is that corresponding to a rigorous mathematical formalism. The "softness" of the human sciences attests to their lack of respect for these three key ideas, which formed a paradigm of simplicity over the course of several centuries. What could be softer, more complex, than a human subject? The exclusion of the subject is therefore a logical consequence. "The death of the human being" coincides with the complete separation of science and culture.

One understands the indignant cries unleashed by the concept of two cultures—scientific and humanist culture—introduced some decades ago by C. P. Snow, a novelist and a scientist.[22] The emperor wore no clothes. The comfort of the owners of the spheres of knowledge was threatened and their conscience was put to the test. Conforming to Snow, science is certainly part of culture, but this scientific culture is completely separated from humanist culture. For him, the two cultures are perceived as antagonists. The split between the two cultures is first of all a split between values. The values of scientists are not the same as the values of humanists. Each world—the scientific world and the humanist world—is hermetically shut on itself.

Of course, the analysis made by Snow was oversimplified and even superficial. In fact, as will be shown in this book, science and culture have always interacted. Visions on what nature is established by science at a given epoch penetrated the imagination of the people living in that epoch. This interaction even leads to great artistic works. Take, as exemplary cases, the interaction between surrealism and modern physics or the interaction of antitheater and quantum physics, cases which will be described in chapters 12 through 14. It is true that, at the institutional level, science was considered as the only way to truth, in contradiction to what really happened on the social level.

Time has passed since 1959, when C. P. Snow forged the expression "the two cultures." The marriage between fundamental science and technology is now accomplished, generating *the technoscientific culture* that drives the huge irrational force of globalization, centered on the economy, which in turn could erase all differences between cultures and even between religions. Part of humanistic culture has already been absorbed in the technoscientific culture. In the face of this new monolithic culture, there is the *spiritual culture*, which is in fact a constellation of a huge variety of cultures, religions, and spiritual communities, sometimes contradictory but still united through a common belief in the two natures of the human being—on one side, the psyche, biological, and psychical nature, and on the other side, the transcendental nature.

Scientists, active participants in the technoscientific culture, have a great responsibility: to avoid the disintegration of spiritual culture resulting from the unbridled development of technoscience, whose probable outcome will be the disappearance of our human species. There are more and more links between technoscientific culture and spiritual culture at the present time. It is only if we question the space between, across, and beyond disciplines that one has a chance to establish links between the two postmodern cultures, integrating both science and wisdom. Transdisciplinarity can offer a methodological foundation for a dialogue between the technoscientific culture and the spiritual culture.[23] This fact is present in many books and articles published in the last decades.

In fact, it is spirituality that links the two cultures. The spiritual dimension was completely absent from C. P. Snow's famous pamphlet.

In any case, the debate created by the concept of "two cultures" has been beneficial, because it has imparted a sense of danger induced by their split: it has exposed the extreme masculinization of our world, with all the dangers this implies for our individual and social life. The expression "two cultures" is not rigorous, but it is a good metaphor for something deep and real.

In recent times, the signs of reconciliation between the two cultures have begun multiplying, above all in the dialogue between science and art, the fundamental axis of a dialogue between scientific culture and humanist culture.

One stage has been passed through by the interdisciplinary rapprochement between science and art. Here, too, the initiatives are numerous and fertile. The acceleration of this rapprochement has an unprecedented rhythm, which is produced before our eyes, thanks to the informatics explosion. There is today a new kind of art, which was created by transferring computer methods to the realm of art. The most spectacular example may be art that uses the incredible information circulating on the Internet as if it were a kind of new "substance." Information rediscovers its original meaning of "in-formation": to create form, ceaselessly changing new forms, arising out of the collective imagination of artists. The interconnectivity of computer networks allows such connections between artists, who come together in real time on the Internet in order to create together, in text, song, and image, a world that arises from somewhere else. This "somewhere else" is found in the inner worlds of artists trying to harmonize, to discover together whatever it is that connects them with creation. These experimental researches constitute the germ of a genuine transdisciplinarity in action.

The encounter between different levels of reality of the subject and of the object (a key notion that will be defined in chapter 8) engenders different levels of representation. Images corresponding to a certain level of representation have a different quality than images associated with another level of representation, because each quality is associated with a certain level of reality. Each level of representation appears like a veritable wall, apparently unassailable because of its relation to the images engendered by another level of representation. These levels of representation of the tangible world are therefore connected with the levels of reality in the being of the creator, the scientist, or the artist. True artistic creation arises in the moment, bridging several levels at the same time, engendering a *transperception*. True scientific creation arises in the moment, bridging several levels of representation at the same time, engendering *transrepresentation*. Transperception permits a global, nondifferentiated understanding of the totality of levels. The surprising similarities between moments of scientific and artistic creation are thus explained, as brilliantly demonstrated by the great mathematician Jacques Hadamard.[24]

If multidisciplinarity and interdisciplinarity reinforce the dialogue between the two artificially antagonistic cultures, transdisciplinarity permits us to envisage the reconciliation of the two artificially antagonistic cultures—the scientific culture and the humanist culture—by virtue of their overlapping within the open unity of cosmodern culture.

THE TRANSCULTURAL AND THE MIRROR OF THE OTHER

To contemplate twentieth-century culture is at once disconcerting, paradoxical, and fascinating.

Since time immemorial, immense treasures of wisdom and knowledge have been accumulating, and still human beings continue to kill each other.

It is true that the treasures of one culture are virtually incommunicable to another. There are even more cultures than there are languages, and the number of languages on our planet is already legion. This is a formidable obstacle to authentic communication and communion between human beings, brought together by our destiny on one and the same Earth. One can translate from one language to another, but translations are forced into merely apparent perfection at the cost of making more or less gross approximations. In the future, one might imagine the appearance of a supercomputer, a kind of universal dictionary capable of furnishing us with a translation of the meaning of the words in one language into the words of any other language, with the same meaning. But a similar translation, be it partial or general, between different cultures is inconceivable, because cultures emerge from *the silence between the words*, and this silence cannot be translated. No matter what their emotional weight, the words of everyday life are primarily addressed to the intellect, the instrument granted to human beings for survival. But cultures emerge from the totality of human beings forming a community in a particular geographic and historical area—all their feelings, hopes, beliefs, and questions.

Prodigious advances in methods of transportation and communication have brought about an intermingling of cultures. Today one finds more Buddhists in California than in Tibet and more computers in Japan than in France. This cultural intermingling is chaotic. The proof: the innumerable difficulties concomitant with the integration of different cultural minorities into various countries in the world. Under what banner could this phantasmagoric integration be performed? No Esperanto, no matter how computerized it might be, can ever promise a translation between cultures. Paradoxically, today everything is open and closed at the same time.

The overwhelming advance of technoscience has served only to deepen the abyss between cultures. The nineteenth-century hope for a single culture in a worldwide society, founded on the happiness brought about by science, crumbled a long time ago. In its stead, we have witnessed an institutional separation between science and culture on the one hand and a cultural fragmentation within each and every culture on the other.

The separation between science and culture engendered the myth of a separation between East and West: the East as the repository of wisdom and knowledge of the human being and the West as the repository of science and knowledge of nature. This separation, at once geographic and spiritual, is artificial, because, as Henry Corbin has stated, the Orient is a facet of the Occident and the Occident is a facet of the Orient.[25] In each human being, the Orient of wisdom (the affective) and the Occident of science (the effective) are potentially reunited. But like all myths, that of the separation of the wisdom of the East and the science of the West is partly true, because modern science really was born in the West. And, in fact, it is true that the spread of the Western lifestyle throughout our planet is associated with the destabilization of traditional cultures. On account of its economic strength, the West has a great responsibility: to avoid the cultural disintegration resulting from the unbridled development of technoscience. Today one has to take the science of the West and the wisdom of the East and only then try to build a new civilization facing all the challenges of the twenty-first century.

Cultural fragmentation is felt in the heart of every culture. The disciplinary big bang has its equivalent in the big bang of cultural modes. As the inevitable result of the loss of frames of reference in an increasingly complex world, one mode of thought is swept away by the next with ever-increasing speed, as Suzy Gablik described in her best seller *Has Modernism Failed?*[26] Before long, through the intervention of computers, the speed of change in cultural modes may reach the speed of light. However, although thanks to scientific methodology, the disciplinary fragmentation within science leads to more or less stable territories, those of cultural modes remain the domain of ephemera. Culture today appears more and more like some kind of monstrous rolling garbage can, in which strange defenses against the terror of nonmeaning proliferate. Of course, as always, the new is hidden in the old, and it is slowly but surely being born. This still formless mixture of the new with the old is fascinating, because beyond all the different cultural modes, a new cultural way of being is taking shape.

In spite of its chaotic appearance, modernity leads to a rapprochement between cultures. With infinitely more intensity than in previous times, modernity brings about a resurgence of the need to unite being with the world. The potential for the birth of a culture of hope is precisely as powerful as the potential for self-destruction engendered by the abyss of nonmeaning.

The multicultural shows that the dialogue between different cultures is enriching, even if its goal is not real communication between cultures. As an example, the study of Chinese and Islamic civilizations is certainly fruitful for deepening the comprehension of European culture. The multicultural helps us discover the face of our own culture in the mirror of another culture.

The intercultural is clearly assisted by the growth of transportation and communication and by economic globalization. A deepening discovery of hitherto badly known or unknown cultures makes unsuspected potentialities burst forth from our own culture. The influence of African art contributed to the appearance of Cubism—an eloquent example. The face of the Other permits us to know our own face better.

Obviously, the multicultural and the intercultural by themselves do not ensure the kind of communication between all cultures that presupposes a universal language founded on shared values, but they certainly constitute important steps toward the act of transcultural communication.

The *transcultural* designates the opening of all cultures to that which cuts through them and transcends them.

The reality of an opening like this is proven, for example, by the research that has been led in Paris for almost half a century by director Peter Brook with his company, Centre International de Créations Théâtrales.[27] The actors are of different nationalities and thus are themselves immersed in different cultures. Nevertheless, during a performance, they reveal qualities that cross and transcend cultures, using a wide range of material from The Mahabharata to The Tempest, from The Conference of the Birds to Carmen. The popular success of these performances in different countries of the world shows us that a transcultural approach can be as accessible to audiences as their own culture.

The perception of that which crosses and transcends cultures is, first of all, an experience that cannot be reduced to the merely theoretical yet is rich with teachings for our own life and for our action in the world. It indicates that no one culture constitutes a privileged place from which one can judge other cultures. Each culture is the actualization of a potentiality of the human being in a specific place on earth and at a specific moment in history. Different places on earth and different moments in history actualize different potentialities of the human being, that is, different cultures. It is the open totality of the human being that constitutes the place-without-place of that which crosses and transcends cultures.

The perception of the transcultural is, first of all, an experience, because it concerns the silence of different actualizations. The space between the levels of reality is the space of this silence. It is the equivalent, in interior space, of that which is called the *quantum vacuum* in exterior space. It is a full silence, structured in levels. There are as many levels of silence as there are correlations between levels of perception and levels of reality. And beyond all these levels of silence, there is another quality of silence, that place-without-place that the poet and philosopher Michel Camus calls "our luminous ignorance."[28] This nucleus of silence appears to us as an unknowable because it is the unfathomable well of knowledge, but this unknowable is luminous because it illumines the very structure of knowledge. The levels of silence and the levels of our luminous ignorance determine our lucidity. If there is a universal language, it goes beyond words, because it concerns the silence between the words and the unfathomable silence that is expressed by each word. Universal language is not a language that can be captured in a dictionary; it is the experience of the totality of our being, reunited at last beyond all its myriad forms. It is, by its very nature, a *translanguage*.

Human beings are the same on the physical level: they are constituted by the same matter, above and beyond their various physical appearances. Human beings are also the same from the biological point of view: the same genes engender different skin colors, different facial expressions, our qualities, and our faults. The transcultural suggests that *human beings are also the same from the spiritual point of view*, beyond the enormous differences that exist between various cultures. The transcultural is expressed by simultaneously reading all of the levels of silence, across a multitude of cultures. "The rest is silence," as in the last words of Hamlet.

It is the Subject who forges *translanguage*, an organic language that captures the spontaneity of the world beyond the infernal chain of abstraction after abstraction. The event of being is just as spontaneous and sudden as a quantum event. It is the sequence of events of being that constitutes true actuality, which, alas, does not receive any attention from our mass media. Yet, these events are what constitute the nucleus of true communication.

That which is found in the center of the transcultural is the problem of *time*. Time is the measure of change of different processes. As a result, time is always thought of in the past or in the future. It is the domain of the Object. In contrast, time that is experienced in the spontaneity of an event of being, the present instant, is unthinkable. As Saint Augustine and Charles Sanders Peirce once observed, the present moment is a point in time in which no thought can occur, no detail can be separated.

The present moment is *living time*. It concerns the Subject; more precisely, it concerns that which links the Subject to the Object. The present instant is, strictly speaking, a non-time, an experience of relation between Subject and Object; thus, it contains potentially within itself the past and the future, the total flow of information and the total flow of consciousness, which cross the levels of reality. The present time is truly the origin of the future and the origin of the past. Different cultures, present and future, develop in the time of history, which is the time of change in the state of being of peoples and of nations. The transcultural concerns the time present in *transhistory*, a notion introduced by Mircea Eliade, which concerns the unthinkable and epiphany.[29]

The *transcultural* is the necessary condition for the existence of culture. The complex plurality of cultures and the open unity of the transcultural coexist in the cosmodern vision. The transcultural is the spearhead of cosmodern culture. Different cultures are the different facets of the human being. In a recent book, Christian Moraru gave multiple references and descriptions of this fact by analyzing the American narrative as manifested in literature.[30] The multicultural allows the interpretation of one culture by another culture, the intercultural permits the fertilization of one culture by another, and the transcultural ensures the translation of one culture into various other cultures, by deciphering meaning that links them and simultaneously goes beyond them.

THE TRANSRELIGIOUS ATTITUDE AND THE PRESENCE OF THE SACRED

The problem of the sacred, understood as the presence of something of irreducibly real in the world, is unavoidable for any rational approach to knowledge. One can deny or affirm the presence of the sacred in the world and in ourselves, but in view of elaborating a coherent discourse on Reality, one is always obliged to refer to it.

The sacred is that which connects. The sacred links, as indicated by the etymological root of the word *religion* (*religare*—"to bind together again"), but such ability is not, in and of itself, an attribute of just one religion.

Mircea Eliade once stated in an interview: "The sacred does not imply belief in God, in gods, or spirits. It is . . . the experience of a reality and the source of the consciousness of existing in the world."[31] The sacred is first of all an experience; it is transmitted by a feeling—the "religious" feeling—of that which links beings and things and, in consequence, induces in the very depths of the human being an absolute respect for others, to whom he or she is linked by their all sharing a common life on one and the same Earth. *Homo religious* always existed and will exist forever, in spite of the contortions of historical events. It is interesting that the notion of the sacred is now the subject of violent debates in academic circles.[32]

The abolition of the sacred led to the abomination of Auschwitz and to 25 million deaths under the Stalinist system. The absolute respect for others has been replaced by the pseudosacralization of a race or of a new human being embodied by dictators elevated to the rank of divinities.

The origin of totalitarianism is found in the abolition of the sacred. Although it is the experience of the irreducibly real, the sacred is actually, as Mircea Eliade repeatedly asserted, an essential element in the structure of consciousness and not simply a stage in the history of consciousness. When this element is violated, disfigured, or mutilated, history becomes criminal. In this context, the etymology of the word *sacred* is highly instructive. This word comes from the Latin *sacer*, which is to say, "that which cannot be touched without soiling" but also, "that which cannot be touched without being soiled." *Sacer* designates the guilty, those who are consecrated to the infernal gods. At the same time, because of its Indo-European root *sak*, *sacred* is linked to *sanctus*, holy. This double meaning of *sacer*—sacred and evil—is the double meaning of history itself, with its contortions and its contradictions that give the impression that history is a tale of madmen.

In 1955, André Malraux was quoted by a Danish newspaper as saying: "With psychoanalysis, our century rediscovered the demons in man—the endeavor which awaits us is now is rediscovering the gods."[33] It is paradoxical and significant that the most desacralized period in history has generated one of the most profound reflections on the question of the sacred. The unavoidable problem of the sacred runs through the work of many diverse twentieth-century thinkers and authors, as well as through that of scientifically minded artists and poets, from masters of thinking to masters of living.

The sacred permits the encounter between the ascending movement and the descending movement of information and consciousness through the levels of reality. This encounter is the irreplaceable condition of our freedom and of our responsibility. In this sense, the sacred appears as the ultimate source of our values. It is the space of unity between time and non-time, causal and acausal.

There is an open unity of questioning in the multiplicity of answers, because the sacred is the question.

One way or another, different religions, as well as agnostic and atheist currents, are defined in terms of the question of the sacred. Experience of the sacred is the source of a *transreligious attitude*. The cosmodern inhabitant of the world is neither religious nor irreligious; he or she is transreligious. It is the transreligious attitude emerging from lived transdisciplinarity that permits us to learn to know and appreciate the specificity of religious and irreligious traditions that are foreign to us, to better perceive the common structures on which they are based, and thus to arrive at a transreligious vision of the world.

The transreligious attitude is not in contradiction with any religious tradition or with any agnostic or atheistic current to the extent that these traditions and currents recognize the presence of the sacred. In fact, the presence of the sacred is our transpresence in the world. If it were widespread, the transreligious attitude would make all religious wars impossible.

At its extreme point, *the transcultural opens onto the transreligious*. Through a curious historical coincidence, the discovery of the Venus of Lespugue occurred in 1922, just two years after the scandal of Brancusi's Princess X, a sculpture banned from the Salon des Indépendants due to accusations of obscenity.[34] Astonished art lovers discovered the shocking resemblance between a Paleolithic sculpture and that of the most innovative artists of the era, who would be later recognized as the founder of modern sculpture. Like the unknown sculptor of Venus de Lespugue, Brancusi tried to make the invisible—the essence of movement—visible. Working in the context of their own cultures, each of these artists attempted to respond to the question of the sacred by making the invisible visible. In spite of the millennia separating the two creators, the forms that issued from their interior beings had a striking resemblance.

The transreligious attitude is not simply a utopian project—it is engraved in the very depths of our being. Through the transcultural, which leads to the transreligious, the war of cultures—an increasingly present menace in our time—has no more reason to be. If the transcultural and transreligious attitudes were to find their proper place in modernity, the war of civilizations could not take place.

Of course, the contrary neoatheistic attitude is also present. The great postmodern thinker George Steiner stresses that the barbarity of the twentieth century is without precedent in human history. Quoting Samuel Beckett ("He doesn't exist, the bastard!") and Bertrand Russell ("It isn't nice of Him not to give us news"), George Steiner expresses his own deep belief in the value of a future atheistic civilization.[35] The fascination of postmodern humanists with technoscience is troubling.

The transcultural characterizes the work of the great Arab poet Adonis in what he calls the *mysticism of art*: a movement toward the hidden face of Reality, a living experience, a perpetual travel toward the heart of the world, a unification of contradictories, the infinity and the unknown as aspiration, freedom from any philosophic or religious system.[36] It is also close to what which the great Christian theologian and philosopher Raimon Panikkar calls the *intrareligious dialogue*: a dialogue that occurs in the heart of any human being.[37]

CHAPTER TWO

CONTEMPORARY PHYSICS AND THE WESTERN TRADITION

He who does not see the camel on top of the minaret, how could he see a hair in the camel's mouth?

—Rûmî, *Fîhi-mâ-fîhi*[1]

TRADITION AND TRADITIONS

An important observation must be made from the very beginning: the word *tradition* (from the Latin word *tradere*, meaning to deliver, to submit) carries within itself a contradiction fraught with consequences.

According to its first familiar and common meaning, the word *tradition* means a way of thinking, of doing or acting, that is inherited from the past and that is thus linked to the words *practice* and *custom*. In this sense, we can talk about an academic tradition, about the tradition of French comedy, or about the Newtonian tradition. In science, tradition is an attempt at mummification, of keeping a particular theory or way of conceiving Reality at all costs. It is very obvious that after this first clarification of the word *tradition*, science is, essentially, antitraditional, because it conceives the research of the unknown, the invention, under the pressure of experimental facts and new theories even more adapted to the description of Reality.

According to its second, less-current meaning (and the only one used in this book), tradition means the ensemble of doctrines and religious or moral practices handed down from one century to another, originally by word of mouth or parable, and also the body of more or less legendary information concerning the past, transmitted, in the beginning, by word of mouth from one generation to another. Following this definition, Tradition incorporates different traditions—Christian, Hebrew, Islamic, Buddhist, Sufi, and so on. In order to avoid any confusion between the two meanings of the word *tradition*, we will write the word with a capital letter when we refer to the second sense.

Therefore, Tradition (in its second meaning) contains essentially the transmission of an ensemble of knowledge of the spiritual evolution of the human being, of our position in the various worlds, and of our relationship with various cosmoses. This body of knowledge is thus inevitably invariant, stable, and permanent, despite the many forms assumed in its transmission and despite the distortions introduced by time and history. Although transmission is most often made by word of mouth, it can nevertheless also be made by writings or works of art or by myths or rites.

Paradoxically, the old idea of an invariant body of knowledge appears to be absolutely unprecedented in the contemporary world, haunted by the cult of personality, whether in politics or in art, in sports or in science.

Traditional knowledge proceeds from immemorial times, but it would be futile to search for an origin of Tradition. In its deep roots, Tradition may be conceived outside geographical space and outside historical time. It is eternally present, here and now, in every man and woman, in a continuous emergence. The origin of Tradition can only be metaphysical. Addressing what is essential to humankind, Tradition turns out to be of immediate actuality.

But Tradition also exists in space and time. Although its content is unique, its form of expression and the language used to express it are of great diversity, under the inevitable influence of history and of the cultural environment. One of the fundamental ideas of Tradition—that of *unity in diversity* and of *diversity through unity*—applies to Tradition itself. The ensemble of a tree's branches does not preclude the existence of a unique trunk. And a tree without branches would be a dead tree.

It would be enough to let ourselves be imbued with the impressions of a Zen monastery in Kyoto and of a baroque church in Bavaria in order to understand the considerable differences in a nevertheless unique search. For a Western scholar, most of the traditional extreme Eastern and African forms of art and spirituality are as far away as Sirius. For a traditional education to have the chance of being truly understood by the modern man or woman, it must necessarily respect language, mentality, culture, and personal history. The fact that science plays an increasingly predominant role in the life of today's man or woman means that any contact with Tradition, even with one's own Tradition, has become impossible.

SCIENCE AND TRADITION: TWO POLES OF A CONTRADICTION

For a typical scientist who makes the effort to be impartial, everything seems to separate science and Tradition.

Traditional knowledge is based on revelation, on contemplation, and on the direct perception of Reality. At the other extreme, scientific knowledge is based on the understanding of Reality by means of the mind and by means of logical and mathematical constructions. Traditional knowledge requires mental silence, eliminating the usual logical associations; scientific knowledge is only possible thanks to the intense activity of the mind.

At the same time, Traditional research attaches a great importance to the body, to sensations and feelings; scientific research excludes the researcher's own body, sensations, feelings, and faith in the field of observation and the formulation of laws. The only instrument belonging to the human body and tolerated by science is the brain and the logical structure inherent in it and common to all researchers. Different experimental instruments are supposed to be equipped with an intrinsic objectivity, an almost absolute independence against the will of the researcher.

Traditional thinking has always asserted that reality is not related to space-time: it *is*. Traditional researchers impose on themselves, through long and bitter toil, the annihilation of their own identity in space-time, with the aim of regaining their true being through dissolution in the unique Reality that includes everything and, in order to be known, does not allow any separation or impurity due to a projection in space or time. At the other extreme, the scientific researcher is obliged to postulate the existence of an objective, separate reality, independent of any observation or measurement, which is necessarily defined in space and time. "There is something like 'the true state' of a physical system, which exists objectively, independently of any observation or measurement, which can be described, in principle, through the medium of physics," Einstein wrote. No one doubts, for example, that at a determined moment, the world's center of gravity would occupy a fixed position, even in the absence of any observer—real or potential. Abandon this view of reality, considered to be logical and purely arbitrary, and you will find yourself in great difficulty of escaping solipsism. For the purposes stated above, I do not blush if I put the concept of a "real state of a system" at the very center of my meditation. But at the same time, Einstein admitted that "this thesis on reality does not have the meaning of a clear statement in itself, because of its *metaphysical* nature; it only has the character specific to a *program*."[2]

Another essential difference between science and Tradition consists in the communicable or incommunicable nature of an experiment. Traditional research has the right to the incommunicable experiment through natural language. The Traditional experiment is unique, complete, surpassing the usual logical categories by far. On the contrary, a scientific experiment is communicable, repeatable. The conditions of a scientific experiment are defined in the most objective manner possible. A scientific experiment can therefore be repeated by any team of researchers equipped with the appropriate scientific devices. The experiment is even considered the supreme court of science. The argument of authority does not exist in science, except as a sociological marginal and transitory phenomenon. A theory, even of the highest aesthetic beauty or formulated by the greatest physicist of his or her time, is dismissed without hesitation if it is in blatant contradiction with the experimental data.

Traditional knowledge therefore claims its *right for ineffectiveness*, in terms of spatial-temporal materiality and in terms of directly observable materiality. When such efficiency is nevertheless achieved, it is regarded by the veritable Tradition as being accidental, even evil (in the sense of separation), as a formidable barrier on the path of spiritual progress. Jacob Boehme enunciates this truth in a precise manner: "The outer body has no power to move the world of light; it only entered this way in the world of light, which made the light to be extinguished inside the man. But the world of dark did not remain less in itself, and the world of light remains motionless inside him, it is as if hidden inside him."[3] Instead, science is primarily interested in the outer body with maximum efficiency in terms of direct materiality. It is only because of this effectiveness that the material life of humankind has been profoundly transformed by the technological applications of technoscientific discoveries.

Finally, a last distinction between science and Tradition concerns language. Concepts such as God, transcendence, and light or dark arouse the disdain of the contemporary scientist, because their integration into the conceptual construction would ruin the very basis of scientific knowledge.

Traditional research addresses the whole person, involving a range of issues infinitely richer than that of modern scientific research, whose purpose is, in fact, the study of that which is outside nature. Tradition is linked to an oral teaching, untranslatable into ordinary language. Is it significant that no Traditional writing ever describes a self-initiation? Saint John of the Cross repeatedly proclaims the appearance of a treatise on the *mystical marriage*, but there has not been found any trace of such a treatise. Farid ud-Din Attar (ca. 1142–ca. 1229) devotes most of his poem *The Conference of the Birds*[4] to the narration of talks between birds—a description of preparations for a trip—but the trip itself and the meeting with the Simorgh only take a few lines.

However, it is true that a gradual modification of scientific language is produced. By exploring increasingly finer areas of Reality, scientific language becomes increasingly more abstract, calling for a more complex mathematical formalism. Thus, scientific language is, in the concrete meaning of the word, untranslatable into ordinary language. One might draw the hasty conclusion that temporary scientific language refers to an incommunicable reality, close to the one probed by Tradition. However, this conclusion would be erroneous. Mathematical formalism can be acquired through an intellectual effort (difficult, certainly, but accessible), which does not imply any moral or spiritual cleansing: *scientific work is not a substitute for a spiritual path*. And the effectiveness of this abstract formalism is given by scientific experiment.

Is there, then, a link between science and Tradition?

In light of the essential differences between science and Tradition, the general opinion is that no rigorous relationship can be established. This conclusion is entirely logical and honorable, and the discussion could stop there. Nevertheless, the conclusion has the disadvantage of being too simplistic and even in contradiction with some fundamental conclusions of modern physics. There is no question of recommending a Hegelian synthesis between science and Tradition, because this could only lead to a hybrid design, which would be fatal for both science and Tradition. The attempt to deal with Tradition by employing the means of science or to deal with science by employing the means of Tradition inexorably leads to destruction and disappointment.

There is another way, which consists in recognizing science and Tradition as two poles of a contradiction and accepting this contradiction, with all its consequences, as a sign of a unique and indivisible reality.[5]

A possible point of contact between science and Tradition is to be found within the limits of the mind itself, limits that can be found in turn by the mind and by science. The characteristic movement of science reaches its own limits, and from there, if these limit areas, border areas, are common to science and Tradition, there could be a leap into another kind of understanding. In other words, contact points can be found in the fundamental axioms of science or in the most general results obtained by science.

A POSSIBLE BRIDGE BETWEEN SCIENCES AND TRADITION: THE RATIONALITY OF THE WORLD

Einstein's famous words are often quoted: "The most incomprehensible thing in the world is that the world could be understood."[6] Paraphrasing these words, one could say that the only irrational issue about the world is its rationality. The playwright Friedrich Dürrenmatt perceives the complexity of Einstein's statement very well when he writes: "The Einsteinian creed about the theory of knowledge is . . . a metaphysical attribute, the first attribute of God that is in line with man's thinking."[7] Indeed, in their everyday practice, scientists continually experience the surprise of seeing agreement between their abstract, logical, mathematical constructions and the experimental facts. The *conformity between human thinking and intelligence hidden in the natural laws* acts as a third term in the humankind-nature relationship. This third term transforms the humankind-nature duality into a ternary entity that exists as a dynamic and inseparable unit. This harmonious, independent term of the humankind-nature relationship explains Einstein's insistence on the role of intuition as a form of immediate knowledge in the genesis of the great scientific discoveries. Forgetting or ignoring this third term seems to be the source of the various contemporary reductionist currents that praise a vulgar, false, and static duality.

Paradoxically, even this creed of the world's rationality is reflected as a constant of Tradition. Traditional thinkers are too often portrayed as apostles of ignorance, of the irrational, forgetting their own writings. However, read what Jacob Boehme wrote: "But you have to take your thought to the spirit and to think that all of nature, with all the powers that make it up—the breadth, the depth, the height, the sky and the earth and everything in it, and what is above the sky—that all these things, I am telling you, are God's body."[8] The respect for Nature, conceived as God's body, implies respect for the intelligence hidden in the laws of Nature. Everything is a sign of this intelligence: "Thus, the study of the world and the study of man will support each other. By studying the universe and its laws, man will study himself and, by studying himself, he will study the universe."[9]

Of course, Traditional experience is incommunicable, but it is very important to notice, in some Traditional thinkers, the need to analyze and explain to others in a meaningful way what has been lived in various stages of the experience. Thus, in 1600, at the age of twenty-five years, Jacob Boehme had a fundamental experience: "In this light my spirit suddenly saw through all things, and recognized God in all creatures, even in herbs and grass, it saw who God was, and how he was, and what his will was. And suddenly in that light my will was set on by a mighty impulse, to describe the being of God."[10]

DESCRIBING GOD'S BEING. . . .

To describe, to analyze, and to explain: this is about a typical Western approach, the approach that underpins contemporary science. The extreme Eastern Tradition prefers a more *direct* enlightening of the indescribable, of the nonanalyzable or of the inexplicable, and even when using techniques of spiritual progress, it places them in the same optic of direct illumination.

Some Traditional thinkers accept contradiction and approximation. An adapted language must inevitably be invented: "Only that, in writing, I must make distinctions for the reader to understand me," says Jacob Boehme,[11] much as how, nowadays, a physicist would try to explain the beauty and truth of quantum laws in his everyday language. Description, analysis, and explanation are parts of a necessary struggle, not without a tragic dimension (in accepting approximation): "What is the use of knowledge if it does not bring you to war? None. It's just as if someone would know about a great treasure, but would not go there to look for it; and while knowing very well where to take it from, dies of hunger despite his knowledge."[12]

A similar approach, *scientifically* extremely interesting, is that of Saint John of the Cross (1542–1591). In the description of his experiences in *Climbing Mount Carmel*,[13] Saint John of the Cross was actually adopting the approach of an experienced contemporary physicist, trying to collect data in order to discover regularities and laws. Of course, he was studying some phenomena of a very special nature and therefore had to invent a new terminology adapted to the new phenomena. Then, by a method of continuous doubt and proof, he tried to describe for others what he had seen. From this point of view, the didactic treatise *Climbing Mount Caramel* seems closer to our concerns today than his poems.

St. John of the Cross was speaking clearly about the limit of a certain logic, of a certain judgment in the face of Reality. In order to be explored, the most subtle levels of Reality require different logics, different judgments, but the world remains material, the world remains rational at every level, because it tolerates description and measurement (and it is interesting to note that the term *reason* comes from the Latin word *ratio*, which implies the meaning of calculation, counting, or proportion). The notion of a *degree of reason* reveals everything that is unclear, vague, or false in the contemporary usage of the word *irrational*. For an admirer of Mallarmé's poetry, Lamartine's poetry may appear irrational, and for a proponent of classical mechanics, the world of quantum mechanics may be perceived as irrational.

Saint John of the Cross indicated, of course, *the night of senses* as a mandatory step during spiritual progress, but at the same time, he saw the sense organs as *windows* and not as walls of the body's prison:

> Thus, the soul that would have rejected the taste of any created things and would have humiliated all its tendencies, could be claimed to have been located in night and darkness; which would have been, somehow, nothing but a full vacuum by contrast with all created objects. The reason for this situation is the fact that the soul . . . is, at the moment when God links it to the body, like a blank or flat sheet on which there was nothing written; and, while leaving aside the knowledge gradually acquired through senses, it receives, naturally, none from elsewhere. . . . Take away from it what it can learn through senses which are like the windows of its prison, and it will certainly no longer be able to get to know anything in a different way.[14]

Thus, there is a possibility of accessing Reality through science, through those windows of man's prison.

Several centuries later, eerily echoing the words of Saint John of the Cross, Einstein, one of the founding fathers of contemporary science, wrote: "The individual feels . . . the marvelous character and the sublime order manifested both in nature and in the world of thinking. Individual existence gives them the impression of a prison and they want to live by owning the perfection of everything there is, in its unity and deep meaning."[15]

Einstein describes instinctively what might be called *the opening of the heart's eye* for overcoming the limits of the rationalizing reason. Access to the various degrees of rationality is conditioned by access to the various *degrees of being*, a value almost completely ignored during our epoch of accelerated fragmentation. "The scientist stops in front of the quantum of action's threshold because he refrains from considering with the heart's eye the non-technical moment when his formalism enters into deadlock," wrote Ludovic de Gaigneron. "Blinded by the success of his method, he cannot conceive its absence as a simple *rational* non-technicality, that only affects the mental structures associated with the sensitive world, nor can he consider the 'non-image' as a strict lack of quantity."[16] Remarkable scientists have proved to be mediocre on the plane of being, but this has not prevented them from making discoveries of the utmost importance. This observation applies to any other area of human activity. Paralyzed when exposed to a degree of being that has nothing in common with the Reality they face, the technicians of the quantitative are unable to access another degree of rationality, thus imperceptibly transforming the rational into the irrational. Yet, science is based on the evolution of rationality. The traditional idea of a necessary harmony between the degree of being and the degree of rationality deserves a long meditation, because it clarifies the very nature of the evolution of rationality.

Confidence in the structuring rationality of the world is the subtle link that connects Traditional thinking and scientific thinking.

MOVEMENT AND DISCONTINUITY: THE ETERNAL GENESIS OF REALITY

Contemporary physics is obsessed with the vision of the world's unity. There was a long road between Einstein's old dream of unification and the current theories about elementary particles, which tend toward unification of all known physical interactions. This tendency was quickly exposed by the recognition of a discontinuity of the physical laws characterizing different levels of Reality: the laws that operate at one level no longer apply at another level. At the same time, the leap into the new laws is not entirely discontinuous. There is a certain continuity, a certain relationship between the different levels, which was fully revealed only at the scale of the infinitely small, the scale of elementary particles. This is why these particles are fascinating, for otherwise they would only be ordinary traces of matter, like the points on a large Seurat canvas, which cannot convey anything of the beauty of the painting as a whole. Paradoxically, the physics of the twentieth century showed that what is happening on the scale of the infinitely small may explain what is happening on the cosmological scale, the scale of the infinitely large. In true self-protection of the secret, this structural asymmetry of the infinitely small scale, toward the invisible, is one of the most disturbing aspects of modern physics. The infinitely small scale is not directly accessible to the human being, but still it is this scale that explains what is happening at the cosmological level. We see planets but we do not see particles.

The palpable, experimental discovery of an invisible scale for the sense organs, the quantum scale, whose laws are quite different from those of the visible scale of our daily lives, was arguably the most important contribution of modern science to human knowledge. The new concept that emerged—that of *levels of Reality*—is counted among those that can create a new vision of the world.[17]

The world of quantum events is quite different from the one we are accustomed to.

The unity of contradictories seems to reign in this new world: quantum entities are both particles and waves. The quantum event is not separable as an object: the new world is that of universal interconnectedness, of relationship, of interaction. Discontinuity and continuity coexist harmoniously (in other words, contradictorily): energy varies by leaps, but our visible world is still that of continuity. The vacuum is filled—it virtually contains all events. The new world is that of perpetual agitation, of annihilation and creation, of a movement with dramatic speeds, far higher than those of our missiles. The energy concentrated at the scale of the small infinity reaches fabulous values, which are barely imaginable on our own scale.

Certainly, the quantum world has its place in the Valley of Astonishment (one of the seven valleys in *The Conference of the Birds*, written by Attar), where contradiction and indetermination pursue the traveler everywhere.

Not surprisingly, the quantum revolution sparked fierce controversies that continue to this day. The greatest physicists of the time were reluctant to accept the significance and consequences of quantum phenomena. Einstein himself, as a true knight of continuity, refused to agree with the conclusions of quantum mechanics until the end of his life. In a letter to Max Born, Einstein wrote: "The quantum theory has brought us many things, but it hardly brings us closer to God's secret."[18]

It is still amazing to note that some traditional thinkers are more in agreement with quantum theory than are some contemporary physicists. We are not talking about the vision of the world in its general form, which is a constant of the traditional philosophy in the East and West. We think rather of the dynamic unity, which places *movement* and *energy* in the center of philosophical reflection.

Jacob Boehme is a precursor of the idea of discontinuity and movement.[19]

At the foundation of the world, of all phenomena, there is a trinitarian law involving three principles. It is present in some traditional philosophies, but it takes on an entirely new dimension for Boehme: "You must return your gaze . . . to the unchanged and sacred trinity that is, *par excellence*, the triumphant, effervescent and active essence, and which, like nature, contains within itself all powers; for it is the eternal Mother Nature."[20] It is about a "triple spirit in which each of them generates the other one."[21] The three principles are inseparable, yet independent, unfathomable and nonlocalizable, yet present inside matter: "Nature and the ternary are not the same. They differ, although the ternary lives in nature, but without being perceived by it; and yet, there is an eternal alliance *between them*."[22]

In order to get the movement, Boehme added a law that puts in movement seven *spirit-origins* or *qualities*: "God is a God of order. . . . Inside him there are, mainly, seven qualities, through which every divine being is active. He looks indistinct in these seven qualities, . . . the divine generation is eternal and serene in its order, and precisely through this law."[23] Every phenomenon is the result of interaction between the seven interchangeable qualities: "The seven *conceptions* are in everything; none of them is the first, none of them is the second, or the third or the last, but each of the seven is the first, the second, the third and the last; . . . divinity is like a wheel composed of seven different wheels located within each other, wherein one could see no beginning and no end."[24]

The dynamism of struggle and cooperation, of annihilation and creation, of eternal movement and endless transformation reigns in the world: "These qualities are brought together, as if they would form one single quality; and yet each quality shakes and buds in its own virtue. Each quality comes out of itself, passes into the others and enhances them while penetrating them, and the other qualities receive something from its will, they feel the liveliness of this quality, what is in it, and they continually join each other."[25]

The particularity of this law is that there is an interval between the first three and the last four qualities, an interval that allows the manifestation of a principle of *discontinuity*. The first three qualities are dominated by the first principle, called *Deus absconditus*, and the last four are a manifestation of *Deus revelatus*. In the interval between the first three and the last four qualities acts a principle called *Fiat*, or "the creative Word," which is really a principle of discontinuity, of life. Without this principle, the world of the seven origin-spirits would be a grave, a dark valley, a virtual inferno.

Finally, there is a mediating principle between the three principles and the seven origin-spirits—*sophia*, or "the contemplation of God": "the emanation is also the joy, as being found in the eternal nothingness, where Father, Son and Holy Spirit see each other and find each other from within, and this is called 'God's Wisdom' (*Sophia*) or the contemplation."[26] In modern language, this would equate with the notion of time; time is the mediator between general laws and what is manifested.

Jacob Boehme's view of Reality is centered on the idea of the unity of contradictories, starting with the "love fight" between the three principles and ending with every created thing: "There is nothing in nature that does not contain the good quality and the bad quality; everything is fretting and living in terms of this double impulse."[27]

Jacob Boehme's universe is a universe of *open* systems, in interconnection and interaction, a universe of the nonseparability: "You do not have . . . to believe that the deity is a being that only inhabits the heaven and that our soul, when it separates from the body, goes to the higher sky, hundreds of miles away. The soul does not need this . . . it can suddenly go to the upper region, and then immediately to the lower region and it is not hindered by anything. Because, in the inmost generation of divine essence, the upper divinity and the lower divinity make up one body. Everything is open."[28] For Boehme, "the true heaven is everywhere, even there where you are and move around at the very moment."[29]

The joint action of the ternary law and of the septenary law produces three worlds contained in one another, an idea anticipating by several centuries the idea of the *three matters* present, five centuries later, in Lupasco's philosophy[30]:

> And so we have to consider a triple being or three worlds, each one being contained in another one. The first world is the world of fire, which derives from the center of the nature. . . . And the second is the world of light that remains free, but which derives from the world of fire. . . . It remains in the fire and the fire does not perceive it. And there is the middle world. . . . The third world is outside, where we remain with the outside body, with the external works and essences that have been created out of *darkness* and, also, out of the world of light.[31]

The structural diagram based on the existence of two fundamental laws, which we have very briefly sketched here, is the major focus of Boehme's works, for example *Aurora: Die Morgenröte im Aufgang* (1612) and *Mysterium Magnum*.[32]

Nikolai Berdiaev (1874–1948) describes Boehme's work by a surprising formula: "First, Boehme conceived of the cosmic life as a passionate struggle, as a movement, as a process, as a perpetual genesis."[33] This eternal genesis is close to what is known about the quantum world, where there is a continuous creation and annihilation of particles, where one can no longer speak about smooth and stable things, and where a continuing fight prevails, associated with a significant increase in energy density. So it is no wonder that Boehme could understand, thanks to a brilliant intuition, this increase in energy density: "The place where the sun exists is such a place that you could choose anywhere on earth. If God wanted to turn on the light from heat, the whole world would be completely similar to the sun. For even this power, where there is sun, is everywhere."[34]

The fact is that traditional thinkers could conceive of the existence of general *isomorphism*, laws that (in the sense that they generate phenomena on all scales, providing a *unity* beyond the infinite variety of the demonstrations on various scales) seem to be of great significance for understanding our modern world, invaded by complexity and fragmentation. Beyond the terminology characteristic of one era or another, it is interesting to clarify the *meaning* of these laws of isomorphism. Only in this way can we discover the convergence between the quantum and systemic view and the traditional view toward a new paradigm.

SCIENTIFIC THINKING AND SYMBOLIC THINKING: ICONS AND THÊMATA

The relationship between Traditional thinking and symbolic thinking is well-known and has been studied in depth. Let us cite as an example the works of René Guénon,[35] which highlighted all that is alive, dynamic, and new in a symbolic reading.

"The symbol is . . . a representation which makes a secret meaning appear, it is the epiphany of a mystery," writes Gilbert Durand.[36] For classical thinking, the symbol can only be illogical, a sign of the occurrence of a new logic—that of the included third, a new logic that claims a new language, radically different from the natural language:

> A *symbol* can never be considered in an ultimate and exclusive sense. To the extent that it expresses the laws of unity in indefinite

diversity, a symbol possesses an indefinite number of features, starting from which the symbol can be taken into account, and it requires from the person who addresses it the ability to see it simultaneously from different viewpoints. Thus, symbols that are translated into the words of ordinary language become harsher, they become hidden and they become very easily "their own opposites," by closing the meaning inside narrow dogmatic frameworks, without leaving even the very relative freedom of a *logical* examination of the subject. The reason is the literal understanding of symbols, the fact that they are only assigned one purpose.[37]

A literal understanding of a symbol turns it into a dead, static concept, without any function and value. In order to understand a symbol, one has to accept a *relativization* of the eye that allows us to grasp the *indefinite* number of aspects of a symbol. This relativization can only be present if the symbol is designed *in movement* and if it is lived by the one who sinks inside its significance. The indefinite number of aspects of a symbol does not mean at all that the symbol is imprecise, vague, or ambiguous. On the contrary: a precise definition implies an inaccuracy of meaning, a *mutilation* of the symbol. Accuracy is present, though, the invariance hidden behind the indefinite multitude of aspects of the symbol. The reading of a symbol, as Alfred Korzybski[38] said, obeys a general *principle of indeterminacy*, one of its special manifestations being Heisenberg's uncertainty relations. The symbol and the logic of the included third are intimately linked.

Classical logic, based on the separation between the different levels of Reality, inevitably implies the gradual *entropy* of language, an extensive language loss over time that decreases its emotional and even intellectual capacity. Things evolved in such a way that language became like a prostitute: one word is taken as another. The different levels of Reality are perceived as failing to communicate with each other. After all, this is exactly the way animals behave. What distinguishes human beings from animals is their *ability to symbolize*. The symbol brings about a gradual decrease in the entropy of language, an increase in *order*, in information, and in comprehension.

As shown, symbolic thinking is operative if Reality has indeed a *multileveled structure*. In scientific language, this would correspond to the existence in nature of *levels of materiality*. The greatest scientific discovery of the twentieth century is the quantum level of materiality, whose laws are clearly distinguished from those that govern Reality on our own scale. The existence of such a quantum scale was not discovered for metaphysical or theological reasons. Science has brought it to light, through the joint effort of increasingly finer experimentation and increasingly more refined mathematics. A crisis of science and even a *dissolution* of Reality have been (and still are being) proclaimed, for the simple reason that these new laws are not the same as the ones that our sense organs have accustomed us to. It is not the dissolution of Reality that is in question but rather the *gradual revelation* of Reality. Abstraction does not estrange us from Reality but brings us closer to it: *abstraction is a component of Reality*, provided that reality does indeed have a ternary structure. The question of the possible link between scientific thinking and symbolic thinking is thus justified.

Apparently, scientific thought and symbolic thinking are incompatible. The world of symbols is perhaps at the antipodes of the world of classical science, but could one make the same statement about contemporary science?

Important steps toward understanding the possible relationship between the world of science and the world of symbols have been made, thanks to Gerald Holton's works.[39] Gerald Holton is a physicist and a famous historian of science. He knows how to highlight the existence of hidden but stable structures in the evolution of scientific ideas. These are what Holton calls *thêmata*, namely the ontological, mostly unconscious assumptions that dominate the thinking of a physicist. These thêmata are hidden even from the person who uses them: they do not appear in the constituted body of science, which only allows logical and mathematical phenomena and sentences. To find them, Holton had to study private documents—the correspondence of physicists, the exchanges of innovative crystallized ideas—*before* moving into the pool of scientific knowledge. Thêmata are concerned with what is more intimate and deeper in the genesis of a new scientific idea. The persistence in time of certain thêmata can capture and shake the faith in novelty at any price, a fashion that has penetrated even into the world of science. It is therefore amazing to note *the restricted number of thêmata* that traverse scientific papers, which are otherwise of a great variety. Holton has reviewed only a few dozen thêmata throughout the history of science, which entitles him to conclude that it is possible that "the phenomenon which, in spite of the developments and the mutations made, assures science the permanent identity which it keeps to a certain extant to be precisely the persistence in time of relatively few thêmata and, also, their circulation, at one time, within the community."[40]

Thêmata are generally presented as double or triple *alternatives*: evolution/involution, continuous/discontinuous, simplicity/complexity, invariance/variance, holism/reductionism, unity/hierarchical structure, constancy/change, and so on. Because of their generality and their persistence over time, thêmata seem to be similar to symbols. With regard to Holton's work, Angèle Kremer-Marietta writes: "before imposing itself on the public as 'science,' the work of scientists is still just an elaboration of symbolizations."[41]

In fact, thêmata are not symbols but rather *facets* of a symbol. A new thêma involves the separation, the opposition of one of the cases of an alternative toward another case (unity *versus* hierarchical structure, for instance). Generalized thêmata are therefore the germs of the cryptofanaticism that is constantly feeding the polemics at the core of science. A symbol as a whole involves the unity of contradictories (*both* unity *and* hierarchical structure, for example).

Holton's satisfaction is understandable when he finds himself in front of what is believed to be a new thêma, as was the case in Bohr's principle of complementarity: "I was struck by the small number of thêmata—in physics at least. . . . The emergence of a new thêma is something exceptional. Complementarity, in 1927, and chirality in the '50s, are the two most recent additions to this category for physics."[42] In fact, the case of Bohr's complementarity is not about a new thêma but about a symbol that carries within itself the unity of continuous-discontinuous and wave-corpuscle contradictions.

An exemplary case of a scientific idea that has all the characteristics of a symbol is the *bootstrap* principle, formulated by Geoffrey Chew in 1959, in particle physics (see Chapter 7). The bootstrap is a dynamic law. According to it, the characteristics and attributes of a physically determined entity are the result of this entity's interactions with other particles: a particle is what it is because all other particles exist at the same time. Therefore, the bootstrap conceives of nature as a global entity, inseparable at a fundamental level.

As a vision of the world's unity and through its consequences on the nature of Reality, the bootstrap principle seems close to Tradition. Was it not Boehme who said, in the *Mysterium pansophicum*: "all together is only one being"[43]? And in *Aurora*, was not he formulating a true principle of the *cosmic* bootstrap when he wrote: "The sun is born and produced by all stars. It is the light extracted from universal nature and, in turn, it shines in the universal nature of this world, where it is linked to other stars, as if all of them together were one single star"[44]?

In an article published in 1968, provocatively titled "Bootstrap: A Scientific Idea?"[45] Chew noted that the idea of the bootstrap, in its most general formulation, is "much older than particle physics." And he continued: "The number of a priori concepts decreases together with the progress of physics, but it seems that science, as we know it, still requires a language based on a number of a priori accepted concepts. So, semantically, the attempt to explain *all* concepts can hardly be called 'scientific.'" Chew concludes: "If at its logical extremes, the bootstrap hypothesis implies that the existence of consciousness, considered together with all other aspects of nature, is necessary to the auto-consistency of everything, such a notion, even if not completely meaningless, is clearly unscientific." Here we find a precise definition of the inexhaustible, irreducible nature of the idea of the bootstrap, which gives it the property of being a symbol.

The contemporary superstring theory, which implies the *dimensions'* bootstrap, seems to be potentially able to link the contradictory aspects of reality: continuity/discontinuity, separability/nonseparability, identity/nonidentity, homogenization/heterogenization, actualization/potentialization. The bootstrap principle thus appears as a principle both structural and organizational, but *hierarchical structure* manifests through the occurrence of the different levels of physical Reality.

A symbol-idea is potentially capable of continuous improvement, of perpetual development in an attempt to find more and more of the undefined richness of the symbol. A theory based on a symbol-idea is thus an open theory. The symbol provides the *permanent* nature of such a theory. An open theory can change, over time, its shape and its mathematical formalism, but its *direction* remains always the same.

A NECESSARY ENCOUNTER

Science and Tradition are distinguished by their nature, by their means, and by their purpose. The only way to understand their interaction is to conceive of them as two poles of one and the same contradiction, as two spokes of one and the same wheel, which, although remaining different, converge toward *the same center*: the human being and our evolution.

A true relationship of contradictory complementarity seems to unite science and Tradition: what Tradition discovers inside the richness of the inner life, science discovers, through isomorphism, in the corporeality of natural systems.

"We could almost define our era as being essentially and foremost 'the kingdom of quantity,'" said René Guénon.[46]

In a way that may seem paradoxical, contemporary science goes against this kingdom of quantity, in spite of the blind applications of its findings, applications that are otherwise beyond its understanding. At the same time, it is clear that *science cannot represent wisdom by itself*, for it deals with human Reality only as a partial aspect. Rather, science helps us to avoid the deadlocks, the ghosts and mirages that arise during the learning process. In turn, Tradition is the memory of the values of the inner life, in a permanent rigor, without which everything may collapse into chaos and destruction. Therefore, the structural convergence between science and Tradition can have an incalculable impact on the present world—or on the future world—through the emergence of a unified, though diverse, image of the world, in which individuals will finally find their place. The conference in

Venice entitled "Science and the Boundaries of Knowledge: The Prologue of Our Cultural Past," organized by UNESCO in collaboration with the Cini Foundation (March 1986), whose organizer I had the honor of being, has meant an important step in this direction.[47] In the Venice Declaration, adopted by seventeen of the world's scientific and cultural personalities (the Nobel Prize winner for physics, Abdus Salam; the Nobel Prize winner for physiology and medicine, Jean Dausset; Maitreyi Devi; Gilbert Durand; Ubiratan d'Ambrosio; Yujiro Nakamura; René Berger; Nicolo Dallaporta; Henry Stapp; and others), the dialogue between science and Tradition was manifested for the first time on an institutional level.[48]

CHAPTER THREE

THE GRANDEUR AND DECADENCE OF SCIENTISM

THE CLASSICAL VISION OF THE WORLD AND THE DEATH OF MAN

Since the beginning of time, the human spirit has been haunted by the idea of laws and order, which give meaning to the universe in which we live and to our own lives. The ancients therefore created the metaphysical, mythological, and metaphorical idea of cosmos. Indeed, they came up with a multidimensional Reality peopled with various entities, from human beings to gods and goddesses, potentially passing through a whole series of intermediaries. These different entities all lived in their own worlds, ruled by their own laws, but everything was linked by common cosmic laws that generated a common cosmic order. Thus, deities could intervene in the affairs of human beings because humans were in the image of the gods and goddesses, and everything had a meaning.

Modern science was born through a violent break with this ancient vision of the world. It was founded on the idea—surprising and revolutionary for that era—of a total separation between the knowing subject and Reality, which was assumed to be completely independent from the subject who observed it. At the same time, modern science was given three fundamental postulates that would extend the quest for law and order on the plane of reason to an extreme degree (see chapter 1).

These postulates can be found, formulated (of course) in the language of the seventeenth century, in the fabulous book of Galileo Galilei: *Dialogue Concerning the Two Chief World Systems*[1] (*Dialogo dei due massimi sistemi del mondo*), published in 1632, which constitutes the founding stone of modernity. In some sense, modernity starts in 1632. The three postulates of Galileo really define what modern science means via its methodology. It is very important to realize that these postulates have remained unchanged until now in spite of the fact that many radically different theories and models traversed several centuries. The postulates were, however, completely fulfilled by only one branch of science: theoretical physics. Perhaps, in the near future, biology will also fulfill Galileo's program.

Thus, conforming to the first principle, Galileo elevated mathematics—an artificial language different from natural languages—to the rank of a common language between God and humanity. The stunning success of classical physics, from Galileo, Kepler, and Newton to Einstein, has confirmed the accuracy of these three postulates. At the same time, these thinkers contributed to the founding of a *paradigm of simplicity*, which became predominant at the beginning of the nineteenth century. During the course of two centuries, classical physics succeeded in building a reassuring, optimistic vision concomitant with the *idea of progress* on the individual and social levels.[2]

In keeping with the evidence furnished by the sense organs, classical physics is founded on the idea of continuity: one cannot pass from one point to another in space or time without passing through all the intermediary points. Moreover, physicists already had at their disposal a mathematical device founded on continuity: the infinitesimal calculus of Leibniz and Newton.

The idea of continuity is intimately connected to a key idea of classical physics: local causality. All physical phenomena necessarily comprise a continuous chain of causes and effects: each cause at a given point corresponds to a nearby effect, and each effect at a given point corresponds to a nearby cause. Thus, two points separated by a distance (perhaps of infinite length) in space and in time are nevertheless linked by a continuous chain of causes and effects: there is no need at all for any direct action from a distance.

The more comprehensive causality of the ancients—for example, that of Aristotle—was reduced to only one of its aspects: local causality. Formal or final causality no longer had a place in classical physics. The cultural and social consequences of such reductionism, justified by the success of classical physics, are incalculable. Even today, most people who do not have a deep knowledge of philosophy consider the equivalence between *causality* and *local causality* as indisputable, so much so that in most cases, the adjective *local* is omitted.

The concept of determinism also had to make its triumphal entry into the history of ideas. The equations of classical physics are such that if one knows the positions and the speed of physical objects at a given moment, one is able to predict their position and their speed at any other moment in time. The laws of classical physics are determinist laws. Physical states are functions of positions and speeds; it follows that if one specifies the initial conditions (the physical state at a given moment in time), one can completely predict the physical state at any other given moment in time.

It is quite evident that the simplicity and aesthetic beauty of such concepts—continuity, local causality, and determinism—whether or not they are operative in nature, have fascinated the greatest minds of these last four centuries.

The final stage was for science to subsume philosophy and ideology by proclaiming physics queen of the sciences. More precisely, everything was reduced to physics; the biological and the physical appeared to be merely evolutionary stages with one and the same foundation: physics. This stage was facilitated by overwhelming advances in physics. Thus, scientistic ideology, which appeared as an ideology of the avant-garde and enjoyed extraordinary development in the nineteenth century, was born.

In the process, some unexpected perspectives disclosed themselves to the human spirit.

If the universe was only a perfectly regulated, perfectly predictable machine, God could be relegated to the status of a simple hypothesis, unnecessary for explaining the functioning of the universe. The universe was suddenly desacralized and its transcendence pushed back into the shadows of the irrational and of superstition. Nature offered herself as a mistress to humankind in order to be penetrated to her depths, dominated, and conquered. Even without succumbing to the temptation to psychoanalyze scientism, one is forced to note that the writings of nineteenth-century scientists about nature abound in the most unbridled sexual allusions. Is it surprising that the femininity of the world has been disregarded, scoffed at, and forgotten in a civilization founded on conquest, domination, and usefulness at any price? As one perverse but inevitable effect, woman is generally condemned to play a minor role in social organization. In the scientistic euphoria of the period, it was quite natural to take for granted the correspondences between economic, social, and historical laws and the laws of nature, as Marx and Engels did. In the final analysis, all Marxist ideas are based on concepts originating within classical physics: continuity, local causality, determinism, objectivity.

If, like nature, history is subjugated to objective and determinist laws, one can make a tabula rasa of the past, through social revolution or some other means. In fact, all that counts is the present, as the initial mechanistic condition. By assigning certain initial socially specific conditions, one can infallibly predict the future of humanity. It suffices for these initial conditions to be imposed in the name of goodness and truth—for example, in the name of liberty, equality, and fraternity—for the ideal society to be built.

This experiment has been made on a global scale, with known results. How many millions of deaths have there been for dogmas? How much suffering in the name of the good and the true? How is it that ideas, so noble in the beginning, can be transformed into their opposite?

On the spiritual level, the consequences of scientism have also been considerable: the only knowledge worthy of its name must be scientific, objective; the only reality worthy of this name must be, of course, objective reality, ruled by objective laws. All knowledge other than scientific is thus cast into the inferno of subjectivity, tolerated at most as a meaningless embellishment or rejected with contempt as a fantasy, an illusion, a regression, or a product of the imagination. Even the word *spirituality* becomes suspect, and its use is practically abandoned.

Objectivity set up as the supreme criterion of truth has one inevitable consequence: the transformation of the subject into an object. The death of the subject, which heralds all other deaths, is the price one has to pay for objective knowledge. As one important French philosopher asserted, the subject becomes just a word in a phrase. . . .[3]

The human being becomes an object—an object of the exploitation of one human being by another; an object of the experiments of ideologies that are proclaimed scientific; and an object of scientific studies, to be dissected, formalized, and manipulated. The human/God becomes a human/object, of which the only result can be self-destruction. The two world massacres of this century, not to mention the multiple local wars—those, too, have produced innumerable corpses—are only the prelude to self-destruction on a global scale.

Or, perhaps, to self-birth.

For, all things considered, besides the immense hope that it has aroused, scientism has bequeathed us one persistent and deeply rooted idea: that of the existence of a single level of reality, in which the only concept of verticality is that of a person standing upright on a planet governed by the law of universal gravity.

MODERN MAHABHARATA-LIKE DRAMA: THE QUANTUM VISION OF THE WORLD

By one of those strange coincidences, of which history alone holds the answer, quantum mechanics, the First World War, and the Russian Revolution all arose at practically the same time: violence and massacres on the level of the visible, the quantum revolution on the level of the invisible. It was as if the death throes of the old world were accompanied by the discrete, barely perceptible appearance of the birth pangs of a new world. The dogmas and ideologies that have ravaged the twentieth century were derived from classical thought, founded on the concepts of classical physics. A new vision of the world demolished the foundations of that old way of thinking.

Precisely at the beginning of the twentieth century, Max Planck was confronted by a problem in physics. Like all such problems, it seemed innocent enough at first. However, in the course of trying to resolve it, Planck made a discovery that, according to his own testimony, provoked a real inner turmoil.[4] The reason for this was that he had unwittingly witnessed the entry of discontinuity into the realm of physics. According to Planck's discovery, energy has a discrete, discontinuous structure. Planck's "quantum," from which the name *quantum mechanics* was derived, revolutionized all of physics and profoundly altered the contemporary vision of the world (see chapter 5).

How can we understand real discontinuity? That is to say, how can we imagine that there is nothing between two points—neither objects, nor atoms, nor molecules, nor particles, just nothing? Here our ordinary imagination experiences an intense vertigo, whereas mathematical language, which is based entirely on another type of imagination, experiences no difficulty whatsoever. Galileo was right: mathematical language is of another nature than everyday human language.

To call continuity into question amounts to calling local causality into question, thus opening a veritable Pandora's Box. The founders of quantum mechanics—Planck, Bohr, Einstein, Pauli, Dirac, Schrödinger, Born, de Broglie, and a few others, who incidentally all had solid philosophical backgrounds—were entirely conscious of the cultural and social implications of their discoveries. This is why they moved with great prudence, avoiding outspoken polemics. Yet, inasmuch as they were scientists, whatever their religious or philosophical convictions, they had to bow to experimental evidence and theoretical self-consistency.

Thus began an extraordinary modern Mahabharata-like drama, which would span the twentieth century until our time.[5]

The formalism of quantum mechanics and, subsequently, that of quantum physics (which took off after World War II with the construction of great particle accelerators) indeed tried to retain the concept of local causality as it appears on the macrophysical level, but from the beginning of quantum mechanics, it was clear that a new type of causality was present at the quantum level, which is the level of the infinitely small and the infinitely brief. According to quantum mechanics, a physical quantity has several possible values, each of which is associated with a specific probability. However, in experimental measurement, one obviously obtains a single result for the physical quantity in question. To abruptly deny the plurality of possible values for this physical, "observable" quantity through the act of measurement might seem obscure, but it clearly indicated the existence of a new type of causality.

Seven decades after the birth of quantum mechanics, the nature of this new type of causality has been clarified, thanks to a rigorous theorem—

Bell's theorem—together with some extremely precise experiments. Thus, a new concept entered physics—that of nonseparability. In the macrophysical world, if two objects interacting in a given moment subsequently separate, they very clearly interact less and less. We think of two lovers compelled to be separated, one in one galaxy, the other in another galaxy. Under normal circumstances, their love would fade and eventually disappear.

In the quantum world, things happen differently. The quantum entities continue to interact no matter what their distance from one another. This appears contrary to our macrophysical laws. Interaction presupposes a connection, a signal, and according to Einstein's theory of relativity, this signal has a limited speed: the speed of light. Do quantum interactions break through this wall of light? Yes, if one insists on protecting local causality, at the cost of doing away with the theory of relativity. No, if one accepts the existence of a new type of causality—global causality, which concerns the system of all physical entities in its entirety. After all, this concept is not so surprising in everyday life. Any community—a family, an enterprise, a nation—is always more than the simple sum of its parts. A mysterious factor of interaction that is not reducible to the properties of different individuals is always present in human communities but is always rejected as the "demon of subjectivity." It behooves us to recognize that we are very far indeed from human nonseparability on our little earth.

In any case, quantum nonseparability casts no doubt on causality itself but only on one of its forms: local causality. It does not cast any doubt on scientific objectivity either but only on one of its forms: classical objectivity, which is based on the belief that there can be no connection present other than the local. The existence of nonlocal correlations enlarges the field of truth, of reality. Quantum nonseparability tells us that in this world, at least at a certain level, there is a coherence, a unity of laws that assure the evolution of the totality of natural systems.

In turn, determinism, another pillar of classical thought, also crumbles.

Quantum entities—quantons—are very different from the objects of classical physics—corpuscles (or particles) and waves. If we want to rely on classical concepts, we must conclude that quantons are at the same time corpuscles and waves or, more precisely, that they are neither corpuscles nor waves. If we want to talk about a wave, it is now a question of talking about a wave of probability, which allows us to calculate the probability of obtaining a final state from a particular initial state.

Quantons are characterized by a certain combination of their physical attributes, such as, for example, their positions and their speeds. The mathematical relations leading to Heisenberg's celebrated uncertainty principle unequivocally indicate that it is impossible to localize a quanton at a specific point in space and in time. In other words, it is impossible to assign

THE GRANDEUR AND DECADENCE OF SCIENTISM 43

a specific trajectory to a quantum particle. The indeterminism that rules on the quantum level is a constituent, fundamental, irreducible indeterminism that signifies neither chance nor imprecision.

Quantum randomness is not *chance*.

The word *chance* corresponds to the word *hazard* in English. In turn, the English word comes from the Arab *az-zahr*, which signifies "play of the dice." It is certainly impossible to localize a quantum particle or to say which specific atom disintegrates at which precise moment, but this by no means signifies that the quantum event is an accidental event, owing to a play of the dice (played by whom?); put very simply, questions like these have no meaning in the quantum world. They have no meaning because they presuppose that there must be a localizable trajectory, continuity, and local causality. In fact, the concept of chance, like that of necessity, is a classical concept. Quantum randomness is both chance and necessity or, more precisely, neither chance nor necessity. Quantum randomness is really a constructive gamble, which has a meaning—that of the construction of our own macrophysical world. A finer material penetrates a grosser material. The two coexist and cooperate in a unity that extends from the quanton to the cosmos.

Indeterminism is by no means *imprecision*; the confusion arises only if the concept of precision is implicitly connected, perhaps unconsciously, to the concepts of localizable trajectories, continuity, and local causality. Until now, the predictions of quantum mechanics have always been verified with great precision by countless experiments, but this precision pertains to the attributes proper to quantum entities, not to those of classical objects. Moreover, even in the classical world, the concept of precision has been radically called into question by chaos theory. One very small imprecision in the initial conditions leads to extremely divergent classical trajectories over the course of time. Chaos is embedded in the heart of determinism. Can planners of all sorts, builders of ideological, economic, and other systems, function in a world that is at once indeterminate and chaotic?

The major cultural impact of the quantum revolution has certainly raised questions for the contemporary philosophical dogma of the existence of a single level of Reality.

CHAPTER FOUR

THE VALLEY OF ASTONISHMENT

The Quantum World

> The movement by which the physical system of the Universe is increasingly moving away from the sensible world . . . is also a movement of ever-increasing proximity to the real world.
>
> —Max Planck, *Initiation to Physics*[1]

In his famous story *The Conference of the Birds*,[2] Farid ud-Din Attar, a Persian poet from the twelfth century, describes the long journey of birds on the lookout for their true king, the Simorgh. The birds cross seven valleys, filled with dangers and marvels. The seventh valley is the *Valley of Astonishment*. In the Valley of Astonishment, there is day and night at the same time: we can see and not see alike, we exist and we do not exist, things are both empty and full. Travelers who hold tightly and at any price to their habits, to what they know, fall prey to discouragement and despair—and the world seems absurd, incoherent, and insane. But if they accept and open to the unknown world, then the new view appears to them in all its harmony and coherence. The same considerations apply fully to those who try to pursue the journey in the quantum world.

ABOUT THE DIFFICULTIES OF THE JOURNEY

The quantum world is essentially (though not exclusively) a world of particles, entities that inhabit the small infinite. In terms of distance, the small infinite means regions of 10^{-13} cm (we could try to conceive such a distance by the following mental operation: take one centimeter and divide it into ten equal parts, then take one of those parts and divide it into ten more parts, and continue to apply the operation thirteen times). This scale is associated with a considerable increase in the energy density and with an increasingly faster movement. For a nonspecialist, the mere evocation of such a scale creates either dizziness, or mistrust, or both at once. The difficulty of visualizing such a small scale is evident, and we cannot refrain from thinking that this is not accessible to our means of investigation.

However, several hundred particles have already been discovered, and the same is expected with regard to several others. Microscopic models of the infinitely small, represented by the large particle accelerators, have been able to penetrate to the depths of the matter, and the interactions between particles have been experimentally studied.

Quantum mechanics, the theory established in 1920–1930, represents the formal basis of modern particle physics, which involves both quantum mechanics and Einstein's relativity theory. Theory and experience go hand in hand in the study of interactions between particles, and this allows us to have a finer understanding of the laws that govern the infinitely small.

But how can we convey the beauty and richness of the laws governing the particle's world? The main difficulty lies in the mathematical nature of these laws. The mathematical tool employed is complex, difficult, and refined, and the appropriation of its use requires years and years of apprenticeship. The untruth of a fast communication of the content of physical laws is therefore obvious. It is precisely because of the mathematical nature of those laws that people without a scientific background are rarely able to express the evolution of contemporary physics. Until such people have assimilated the characteristics of higher physics, they cannot factor in to any discussion of it that it evolves, it changes; new laws are discovered, more refined and more subtle.

Another important obstacle to a good scientific popularization is related to the inappropriate images employed with the aim of illustrating quantum laws in a simple manner. Most images do nothing but transpose the laws corresponding to a scale of fundamentally different nature in classical, macroscopic language. In a way, it is like trying to dress an adolescent in baby clothes. On this note, we are in the heart of the matter: *in the quantum world, we cannot create anything new from what is old.* If we forget this fact, we have pursued the opposite goal. By wanting to express quantum laws in classical language, we distort them to the point of making them unrecognizable, we empty them of their meaning, and we create the illusion of understanding and the most extravagant of chimeras.

Acquiring some basic knowledge of mathematics and physics would be a suitable preparation for travelling in the quantum world. But it is absurd to impose such a precondition on knowledge enthusiasts.

A way out of all these difficulties exists. The most general results in physics involve a sort of globalizing simplicity, an aesthetic beauty that not only addresses the mind but also the intuition and sensibility by engaging the deeper layers of fantasy. Indeed, this is exactly where the hidden engine of great discoveries can be found. The mathematical foundations underlying physical laws are untranslatable into ordinary language, but the most general results, obtained by using mathematical tools, can be understood and appreciated by nonspecialists.

Even this way is not without its dangers and must be followed with extreme rigor. The excessive poeticizing of scientific facts leads to a superficial and distorted view of science. It is precisely the absence of rigor that seems to be the source of the Orientalizing fashion. Of course, the Oriental tradition is one of great richness, and it would be absurd to put it in opposition to the Occidental tradition—universal laws do not occur more in the East than in the West. It is highly symptomatic that works addressing the similarities between Oriental philosophy and modern physics practically never address an obvious problem: if Oriental outlooks are truly that similar to those underlying modern physics, why has modern physics been created in the West? In my essay about Jacob Boehme,[3] I have addressed this issue, and I have provided some arguments in favor of a deep connection between the Western tradition and modern physics.

The existing popularizing books and papers discuss extensively the so-called quantum paradoxes (the Einstein-Podolsky-Rosen paradox; the "Schrödinger's cat" paradox; and so on). In fact, they are false paradoxes, because they only reveal contradictions related to the natural, habitual language characteristic of classical realism; if these paradoxes are instructive when they seek to demonstrate the incompatibility between classical realism and quantum realism, they become useless in the context of a first approach to quantum ideas.

PLANCK, DISCONTINUITY, AND THE QUANTUM REVOLUTION

The twentieth century began with a revolution about which humans in the future will probably be talking a lot more than about other revolutions that have crossed the century that just ended. On December 14, 1900, Max Planck presented his work on black body radiation (a *black body* is a body that completely absorbs electromagnetic radiation) at the German Physical Society in Berlin. "After several weeks, which have certainly been full of the bitterest labor in my life," Planck writes, "a flash shone in the darkness where I was struggling, and unexpected perspectives opened to me."[4]

The "flash in the dark" was a concept—the elementary quantum of action (*action* is a physical quantity corresponding to an energy multiplied by a time)—that would revolutionize all of physics and would deeply change our view of the world. This quantum is expressed by a universal constant (Planck's constant) that has a clearly defined numerical value and occurs through integer multiples.

Planck's quantum introduces a discrete, discontinuous structure of the energy. The old, omnipotent construct of continuity was shattered and, together with it, the very foundation of what is regarded as reality. As Planck says, "this quantum represents . . . something entirely new, previously unsuspected, and that seemed destined to revolutionize thinking in physics based on the very concept of continuity, inherent to any causal relationships, after the discovery of the infinitesimal calculus by Leibniz and Newton."[5]

Indeed, continuity is a constant of human thought. It is probably based on the evidence supplied by our sense organs—continuity of our own body, continuity of the environment, continuity of memory. It holds the visible range, that of the constant shape or what evolves in a constant manner; it holds the object itself. Death, natural disasters, and mutations have been considered rather as manifestations of accident, chance, or unfathomable mystery.

In an attempt to understand the whole strangeness of the idea of discontinuity, let us imagine a bird that jumps from branch to branch in a tree, without passing through any intermediate point: it is as if the bird would materialize instantly on one branch or another. Certainly, our ordinary imagination blocks such a possibility. But mathematics can address this sort of situation very precisely.

It is true that traditional thinking, whether Eastern or Western, has always stressed the importance of discontinuity. For instance, in a very instructive paper, Lilian Silburn emphasizes the role of discontinuity in Indian philosophical thinking.[6] But in order to develop, science needs a mathematical device, and it so happened that Newton and Leibniz discovered such a measure based on continuity. One could continue the endless epilogue about the possible development of science if, a few centuries ago, other mathematics had been discovered instead of the infinitesimal calculus of Leibniz and Newton—mathematics adapted to the study of discontinuity—but history is what it is. For centuries, scientific thought has been fed up with the idea of continuity.

No wonder that the idea of discontinuity has produced violent polemics and intense opposition from physicists, even long after its discovery by Planck. For instance, in 1926, during a visit to Copenhagen, Schrödinger, who was one of the founders of quantum mechanics, violently opposed Bohr, who was advocating the idea of discontinuity and the concept of a *quantum jump*: "If I need to join this evil quantum leap," Schrödinger exclaimed, "I will forever regret that I have been involved in its formulation."[7]

It is not difficult to understand the deep connection between the notion of continuity and that of local causality. For classical thinking, the universe was a veritable machine composed of distinct parts that combined with one another according to well-defined laws but maintained the identity of distinct parts. Objects were considered to be subject to the laws of motion, and if we knew the initial conditions (the physical state at some point in time), we would be able to predict the physical state at any other moment in time. "If determinism is true after all," says Weizsäcker, "inferring facts from documents can be replaced by a strict deduction of past and future events, starting from today; this is precisely the way that classical realism conceived the world."[8]

Planck's discovery led the way for a deep foray into the question of classical realism, with all its company of sacrosanct concepts (continuity, local causality, determinism, objectivity, and so on). Nothing, however, predicted that Max Planck (1858–1947) would be the initiator of a great revolution. Emilio Segré, Nobel laureate in physics, rightly called him "the unwilling revolutionary."[9]

Having received a severe Protestant education, Planck was by nature conservative and respected the values of his time. In fact, he had, for years, obstinately attempted to explain his discovery through classical physics, but he had to bow to the evidence that classical physics was not able to explain this discovery. His *Autobiographie scientifique* (*Scientific Autobiography*) allows us to discern the entire complexity of his intimate process of enlightenment: "My vain attempts to adapt the elementary quantum of action to classical physics in one way or another continued for a number of years and they cost me much effort. Many colleagues considered that there was something in there that looked like a tragedy. But in this regard I have a different view. The perfect light that I perceived has undoubtedly enriched me beyond comparison, because then I knew with all certainty, with regard to the elementary quantum of action, that it played a much more important role in physics than I had foreseen in the beginning."[10]

A renowned physicist who received the Nobel Prize for physics in 1918, Planck was also a great philosopher of nature. His reflections on the relationship between science and religion, on the role of irrationality in scientific knowledge, and on the role of abstraction in dealing with the real remain of great current interest: "What characterizes the evolution of physics," Planck wrote in 1925, "is a trend toward unity, and unification is realized especially under the sign of a certain liberation of physics from its anthropomorphic elements, particularly from the relationships that bring it back to what is specific in the perceptions of our bodily senses."[11]

THE PARTICLE AND QUANTUM SPONTANEITY

Classical physics admits two well-differentiated types of objects: *corpuscles* and *waves*. Classical corpuscles are discrete entities, distinctly localized in space, and characterized, from a dynamic point of view, by energy and quantity of motion. One could easily visualize the particles as balls that are continuously moving in space and time while describing a well-defined path.

With regard to waves, they were conceived of as occupying all of space, also in a continuous manner. A wave phenomenon can be described as a superposition of periodical waves, characterized by a spatial period (the wavelength) and a time period. Similarly, a wave can be described by frequencies: a frequency of vibration (the reverse of a time period) and a certain number of waves (the reverse of a wavelength). Waves can also be viewed in a simple way.

Quantum mechanics produces a total reversal of this representation. Quantum particles are particles *and* waves simultaneously. Their dynamic characteristics are established by Einstein-Planck and de Broglie's formulas: energy is proportional to temporal frequency (the Einstein-Planck formula), and the amount of motion is proportional to the number of waves (de Broglie's formula), the proportionality factor being, in both cases, precisely Planck's *elementary quantum of action*.

This representation of a quantum particle challenges any representation by shapes found in space and time, because it is manifestly impossible to mentally represent (other than by mathematical equations) something that is both particle and wave. At the same time, energy is quantified and varies in leaps, discontinuously. The concepts of continuity and discontinuity harmonize, not through an effort of philosophical or metaphysical speculation but through the intervention of the supreme judge of physicists, which is nature itself (in other words, under the pressure of experimental, repeatable, and verifiable facts).

It should be understood that the quantum particle is an entirely new entity that is irreducible to classical representations: the quantum particle is not a simple juxtaposition of a particle and a wave. Numerous experiences, some very recent, have shown that fundamental fact unequivocally. It is true that the results of an experience may nevertheless be analyzed in terms of a particle or in terms of a wave, but these classical concepts prove to be approximate, valid only on a macroscopic scale. Those two aspects coexist inside an experience.

In this respect (and only in this respect), can we understand the quantum particle as a unity of contradictions? It would be fairer to say that a quantum particle is neither corpuscle nor wave: the unity of contradictions is more than the simple sum of its classical, contradictory (in the classical representation), and approximate (with regard to the quantum representation) components.

Now we can understand the absurdity of a representation through images of quantum particles. In the twenty-first century, quantum particles can no longer be represented by a ball that collides with another ball, as they are in most of the popularized works, because that would be the same as the elimination of all meaning for quantum phenomena.

This is also how one can understand the reasons for the passionate atmosphere that has been ruling (and still does) the debates on the interpretation of quantum mechanics: the confrontation between two physical *scales*, corresponding to two different levels of materiality and Reality—the macroscopic scale, including us as observers, and the microscopic scale, the abode of entirely new laws—is the source of this passion.

There have been numerous attempts to reduce quantum phenomena either to their wavy aspect or to their corpuscular aspect, but all of them have failed.

Classical concepts received the finishing stroke from Max Born in 1926. Schrödinger, a strong advocate of continuity, argued that the quantum particle is not a corpuscle but a wave of matter. He brought as an argument his own equation (now famous), which was in remarkable agreement with the experimental data. It was precisely at this moment that Born made his essential observation. He showed that Schrödinger's wave function could be understood as a correlate of the probability of finding a quantum particle (in this case, the electron) in a particular point in space. More specifically, the wave is not a wave of matter, as Schrödinger believed, but a *wave of probability*. The amplitude of this wave is one of probability, and its square gives us the probability of achieving a final state, starting from a certain initial state. These amplitudes of probability have an additive nature; they can be juxtaposed, which expresses the content of the principle of quantum superposition. In contrast, the probability, which is the square of the amplitude, obviously does not possess this additive property (characterizing only the classical corpuscles). The probabilistic nature of quantum events is now unanimously accepted, even if its interpretation still allows room for controversial debates. This probabilistic nature does not mean in the least that "God plays dice with the universe," as is often said when one of Einstein's expressions is uncritically assumed to be true. This is certainly not the sign of some kind of ignorance or inaccuracy in our knowledge of the quantum world—quantum mechanics makes accurate and thorough predictions, fully verifiable by experimental data. Ignorance and lack of precision are only the illusions of the one who forces, at any price, description of quantum phenomena in the framework of classical realism. It is equally wrong to assert that quantum physics only does statistical predictions on a set of particles; it successfully describes individual systems as well.

The probabilistic nature of quantum events is the sign of occurrence of a new property, *quantum spontaneity*. This is linked to the existence of a quantum freedom irreducible to the canons of classical determinism.

HEISENBERG'S RELATIONS AND THE FAILURE OF CLASSIC DETERMINISM

Heisenberg's famous relations shed surprising light on the dynamics of quantum particles. A quantum particle, neither corpuscle nor wave, is characterized by a certain extension of its physical traits: energy, duration, location, and quantity of motion (which is related to the mass and the velocity of the particle).

Heisenberg's relations show that the product between the extension in the quantity of motion of a quantum event and its spatial extension or the product between its extension in energy and its temporal extension must be superior to the elementary quantum of action. In contrast, for a classical corpuscle, the products are zero.

Can a quantum particle be interpreted as the equivalent of a classical corpuscle? If, for instance, a precise, punctual, spatial localization of the quantum event is required, Heisenberg's relations show that the extension in the amount of motion becomes infinite, and if a precise, punctual, temporal location is required, the extension in energy must be infinite. Absurd question, absurd answer. The key to understanding these results lies in the value of the finite, non-zero Planck's constant.

There is no need for highly sophisticated mathematics or physics to understand that this result means *the impossibility of a precise localization in space-time of a quantum event*. The identity concept of a classical particle (identity defined in relation to the particle itself, as a separate part of the whole) is no more valid. According to the beautiful expression of David Finkelstein, "The indeterminacy of quantum phenomena is a special case of the impossibility of self-discovery in the finite systems."[12]

Very often, Heisenberg's relations are called "relations of uncertainty." It is really very bad terminology, even if it does have a historical justification. Using the term *uncertainty* seems to indicate some imprecision related to our measurement devices or to our knowledge of quantum events. That is absolutely false. Quantum events have, by definition, a specific extension in space or time. The illusion of uncertainty, of imprecision, comes, again, from the classical interpretation of quantum events. If there are uncertainty and imprecision, they concern precisely the classical concepts.

Heisenberg's relations should rather be called *certainty relations*. They determine the size of particle accelerators and the stability of atoms. Through experimentation that increases our knowledge of the energy spectrum, they also help us to determine a particle's lifetime. Heisenberg's relations show, in particular, the existence of that fascinating quantum effect that is the increase in the energy density toward the infinitely small. The amount of movement (and therefore energy) must be greater than the ratio of Planck's constant and the spatial extension of the quantum event. *The smaller the region surveyed, the higher the corresponding energy.*

Once Heisenberg's relations are established, Laplace's dream of an absolute determinism crumbles: spontaneity and freedom are integral parts of physical reality.

THE MULTIPLICITY OF QUANTUM VALUES AND THE ROLE OF OBSERVATION

The theoretical description of a physical system involves the description of the *physical state* of the system and the description of its properties, called *physical observables*. Physical states are characterized by *wave functions*. They can be represented by vectors in an abstract space (Hilbert's space) of any number of dimensions. For an isolated system, there is a new kind of determinism: knowing a state at a certain time allows us to know that state at a completely different time. This determinism is itself abolished in the case of the nonisolated systems, which interact.

Physical observables were represented in classical mechanics by functions of positions and speeds: once the positions and speeds in question were defined, the observable was completely determined by a number (the numerical value of the corresponding function). The situation in quantum mechanics is radically different. The observables are described by mathematical entities other than their functions—the operators. The operators are mathematical entities that act on certain functions (*eigenfunctions*), reproducing them, but multiplied by numbers (*eigenvalues*).

The operators have very different properties from those of pure numbers. When multiplying two numbers, say 2 and 2, no matter the order in which the multiplication is performed (the first number multiplied by the second or the second by the first), the result is always the same: $2 \times 2 = 4$. But when multiplying two operators, the situation is very different: the result may depend on the order in which the multiplication is performed (the two operators do not "switch"): 2×2 is different from 4! The mathematical understanding of Heisenberg's relations is to be found in the *noncommutability* of the operators associated with the position and quantity of motion of a quantum particle.

Each operator is assigned a set of proper values—not one, but several proper values, each having a certain probability of manifestation. The physical state can be said to correspond to an overlap, to a *wave packet*. Therefore, a measurement can determine, in principle, different results. Obviously, however, only one of those results will be effectively achieved in an experimental measurement. In other words, the act of measurement cancels the plurality of possible values of the physical observable concerned. This process is called the *wave packet reduction*. The deep nature of the reduction is very obscure, because the fundamental laws of quantum mechanics seem to cease to act in the measurement process. Before the measurement, the system is described by a wave packet, but after the measurement, the system is assumed to be in a state characterized by a well-determined value of the target observable. Thus, there is a discontinuity in the evolution of the state—this evolution ceases to be deterministic in the quantum meaning of the word.

From such considerations, one cannot draw the extreme conclusion that "the reality is created by the observer" or even the more nuanced conclusion that "we cannot talk about a reality that is independent of the measurement process" (the Copenhagen interpretation). It is somewhat strange to meet, among the supporters of the Copenhagen interpretation (which has been adopted almost unanimously by physicists), a thinker of the stature of Niels Bohr. His intimate writings (correspondence, discussions, and so on) reveal that he participated against his will in the development of such a view. The focus on the classical, macroscopic use of language in the description of experimental results gives an undoubtedly operational, neopositivist character to the Copenhagen interpretation. In the face of the overall novelty of quantum phenomena, it is about a compromise, which had a certain historical justification: the Copenhagen interpretation allowed physicists not to devote every effort to problems of interpretation; instead, they were able to focus on getting concrete, technical results that led to the extraordinary development of quantum physics.

But today, maintaining an uncritical attitude toward this interpretation is an obstacle to understanding quantum physics. It is symptomatic that only a tiny minority of physicists are still interested in these questions of the interpretation of quantum physics. As Bernard d'Espagnat stressed, "the considerable successes of current particle physics [have] greatly increased the generality of the formalism of the relativistic and quantum physics and our confidence in this theory. But, contrary to what happened in the '20s, they have not brought anything new out of what might be called a conceptual revolution. As a consequence, the problems of interpretation, that emerged during the Bohr-Einstein debate, and that were then only partially or completely unresolved, still represent real problems nowadays."[13]

QUANTUM VACUUM: A FULL VACUUM

According to classical belief, the vacuum is truly *empty*. It is like a receptacle that contains nothing: you have to bring objects and their interactions from the outside in order to populate this gap. That concept has an undeniably psychological basis. When we look around, we have the feeling of an empty space populated by objects and beings. The vacuum seems to prevail on the planetary, galactic, or intergalactic scale and even on the atomic scale. Aren't so-called solid bodies formed primarily from vacuum? The substance at the atomic scale is concentrated in very small regions, much smaller than the distances that separate these concentrations. Quantum physics tells us that everything is appearance, an illusion created by our own scale. As we penetrate a smaller region of space, we find a growing turmoil, a sign of a continuous movement.

The key to understanding this paradoxical situation is again provided by Heisenberg's uncertainty principle. A tiny region of space corresponds, by definition, to a very short time, and therefore, according to Heisenberg's principle, to a wide spectrum of energies. As a result, the energy conservation law can be violated for very short periods of time: everything happens as if the quanta of matter were created from nothing. In particular, the *quantum fluctuations* of the vacuum cause the sudden occurrence of particle pairs—virtual antiparticles that annihilate each other afterwards; this process takes place within a very short time. The void seems to randomly fluctuate between being and not being.

Everything is vibration: any single point in the universe that would be inert, immobile, unoccupied by motion cannot be conceived, according to quantum physics. Of course, during long time intervals, the energy conservation law is carefully observed: the almost imperceptible violations of this law on the small scale compensate each other on the larger scale.

This way, we can understand why the void appears as a void on our scale, illusion, created by the long distances and the long time intervals. *On a quantum scale, the vacuum is full*: it is the abode of the spontaneous creation and annihilation of particles and antiparticles.

Quantum fluctuations of the vacuum are smaller at the atomic level but noticeable. Numerous experiments have revealed the existence of such vacuum fluctuations. For example, an electron-positron pairing may suddenly occur and be annihilated, and the void polarization thus created can easily disrupt the orbit of an electron that revolves around an atomic nucleus. Willis Lamb's famous experiment (1947) demonstrated the existence of these changes in the electronic orbit of the hydrogen atom.

The quantum particles created by vacuum fluctuations are virtual particles; they do not materialize into real, observable particles. Real particles have some mass, and therefore, according to the theory of relativity, they need a certain energy in order to materialize. Virtual quanta generally correspond to a lower energy than those *energetic thresholds*, leading to the emergence of real particles. By delivering energy to the quantum vacuum, we can help materialize all its potentialities. This is exactly what we do when building particle accelerators. The energy provided in this way determines the transformation of virtual quantum particles into real quantum particles. Particles that have a small mass (for example, the electron and the positron, which have a mass 1,840 times smaller than the proton) have a greater chance of materializing quickly. In contrast, heavier particles require the construction of larger and larger accelerators, capable of providing a more and more important energy. Thus emerges an interesting *dialectic of the visible and the invisible*: in order to detect tiny particles, we have to build those towering cathedrals that are, for physicists, modern particle accelerators. For example, in order to find W and Z, very heavy particles (each with a mass almost one hundred times larger than that of a proton), very high energies had to be reached at the proton-antiproton accelerator in the Centre Européen de Recherches Nucléaires (the European Center of Nuclear Research—CERN) in Geneva.

The full quantum vacuum potentially contains within itself *all* the particles that have been observed or that have not yet been. This remark leads us to raise the following question: what do we understand, in fact, when we speak about a particle *existing in nature*? Our world, the natural world, seems to be built in an extremely economical manner: the proton, the neutron, and the electron are sufficient to build up our visible universe almost entirely. Yet, we have discovered hundreds of hadrons and also a few leptons and electroweak bosons, which do not exist naturally in our universe. It was human beings who extracted them from nowhere while building accelerators and other experimental devices. In this regard, human beings are participants in a reality that encompasses humans, particles, and universe.

We can also analyze the true meaning of *the experimental confirmation* of a theory. The quantum vacuum contains infinite possibilities of experimental life. During our experiments, we necessarily make a choice, a selection of experimental facts that we deem significant. One might think that physics is inevitably engaged in a social construction of Reality. In fact, the role of the human being is to find and give a sense to reality.

The quantum vacuum is a wonderful facet of reality that shows us that we should not stop at the illusions created by our own scale. Quanta—vibrations, whether real or virtual—are everywhere. The vacuum is full of vibrations. It contains, potentially, all of reality. The whole universe might have been extracted from nowhere by a gigantic vacuum fluctuation—the big bang.

QUANTUM NONSEPARABILITY

Nonseparability has always characterized the quantum world, but this property remains, to a certain extent, rather qualitative and related to issues of interpretation. In 1964, Bell was able to establish one of the most important results of quantum physics. He developed a theorem, expressed in the form of certain inequalities, which allows a precise, rigorous, and experimental testing of the most general ideas on local causality and separability.[14]

A detailed analysis of Bell's inequalities and their implications was made for a wide audience by Bernard d'Espagnat in his book À la Recherche du Réel (In Search of Reality).[15]

Let us suppose two experimenters, separated in space-time, who have complete freedom to elect the conditions of their own experiences. The results of one of the experimenters should be—according to ideas of local causality and separability—completely independent of the results of the other experimenter. The very notion of scientific objectivity seems to claim the absence of nonlocal connections, although Bell's theorem shows clearly that the predictions of quantum mechanics cannot be reconciled with the independence of the results.

Bell's inequalities are not philosophical inequalities: they do not involve an a posteriori interpretation, and they pass the exam of the experiment. The overwhelming majority of experimental results is consistent with the predictions of quantum mechanics. In 1980, the French physicist Alain Aspect and his collaborators did a great experiment that brilliantly confirmed the experimental existence of nonlocal correlations. Aspect's experiment revealed the existence of nonlocal correlations on distances of a meter (which is huge compared to 10^{-13} cm). Technical advances have allowed us, fifteen years later, to reach more than 20 km, thanks to an experiment made in Switzerland with an optical fiber cable linking Geneva and Lyon.[16]

Therefore, quantum nonseparability manifests itself, in our own macroscopic world, at distances of nineteen orders of magnitude higher than those of the quantum world. Can we see in those nonlocal correlations the intervention of the irrational, the violation of causality, and the bankruptcy of scientific objectivity? Certainly not. What is in question here is a special, limited form of causality—local causality—not causality itself. Local causality is a special, limited form of scientific objectivity, based on the notion of separate objects, independent of each other. The world remains causal, the world remains rational. But our vision of the world has to change, and our minds have to adjust to a higher degree of finesse. Here is a more detailed description of quantum nonseparability: "if the concept of a reality independent of man is regarded as having a meaning, then such a reality is necessarily inseparable. By the word *inseparable*, we must understand that if, in this reality, we want to imagine some parts located in space, then we must believe that if some spaces have interacted in different ways at a time when they were close together, they will continue to interact, whatever their mutual disposal, by momentary influence."[17]

All the implications of nonseparability arise as a sign of the unity of the quantum world. Moreover, all the parts of our universe were in contact at the beginning of the big bang, when quantum processes must have been preponderate. *Is it likely that the unity of the quantum world implies a unity of the whole universe?*

Quantum nonseparability has a far more subtle and interesting significance than classic mental habits try to make us believe. It tells us that there is in the world, at least to a certain scale, some coherence, some unity, laws that ensure the development of all natural systems.

The whole problem is to properly understand the meaning of the words *instant* and *influence*, because they can unleash the most dangerous confusion. *Influence* does not mean that a signal from a transmitter can be instantly transmitted to a particular receiver. This fundamental fact can be understood quite easily. Nonseparability concerns the whole system, not a separate system. To aim to handle a transmitter in order to send a message basically means that we consider the transmitter as a separate, isolated system, and therefore the nonseparability conditions have been removed. Quantum spontaneity is fortunately incompatible with the macroscopic handling of nonseparability. When someone raises his or her right arm, he or she does not influence what is happening on Sirius. The quantum world is fragile: its interaction with the environment—and so with our own world—abolishes the quantum laws. Coherence (expressed as the quantum principle of superposition) is transformed into incoherence, a technical term that describes a new field of physics. It is this incoherence that enables us to understand the paradoxical coexistence of the quantum world and the macrophysical world.[18]

∽

Is the quantum world really an abstract world, far from us? In fact, it is in us and with us, in our everyday life.

Quantum effects have even some influence on our biological and mental life. The physicist Heinz Pagels gives as an example: "the random combination of DNA molecules when conceiving a child, for which the quantum characteristics of the chemical bond play a role."[19] Also, quantum effects certainly play a specific role in the functioning of our brain and our consciousness.[20]

And what if, ultimately, our own world was the real Valley of Astonishment?

CHAPTER FIVE

THE ENDLESS ROUTE OF THE UNIFICATION OF THE WORLD

IS A SINGLE ENERGY THE SOURCE OF THE WORLD'S DIVERSITY?

From the point of view of monotheist religions, God is One. But what can we say of nature? Is nature also One?

The four known physical interactions are very different from each other. Taking into account, for example, the *intensity* of the interaction: the strongest, as its name shows, is the *strong interaction* (which is exercised between hadrons). The *electromagnetic interaction* (acting on electrically charged particles) is a hundred times weaker than the strong interaction. The *weak interaction* (which is exercised between leptons and hadrons) is really weak: it is one hundred thousand times less intense than the strong interaction. As for the *gravitational interaction* (which acts on all particles), it is tremendously less intense: it is 10 to power 39 times less than the strong force (10^{-39}).

The same diversity appears when you take into account the *range* of the interaction. Gravitational interaction and electromagnetic interaction have infinite range. They play their role on a macroscopic scale, which explains why they were the first to be recognized as physical interactions. In contrast, strong and weak interactions have a very small range of action. A centimeter must be divided 13 or 14 times by 10 (10^{-13}) in order to get the distance at which the strong interaction acts. The scale of the weak interaction is even smaller (much smaller) than that of the strong interaction.

Despite this great diversity, physicists believe today that all interactions can be *unified* into one and the same theory. "All forms of energy are essentially reduced to one," said the Pakistani Nobel laureate in physics Abdus Salam (1926–1996) in an interview in 1982. "The forms that have occurred, those of the strong, the weak, and the gravitational electromagnetic interactions, are all manifestations of a single form of energy."[1] Was Salam saying that because he was a Muslim or because science is saying that?

Such a possibility of unification could occur for two reasons, one experimental and one theoretical.

First, the theory of the unification of weak and electromagnetic interactions (therefore a partial unification), formulated in 1967 by Salam and Weinberg (and which went almost unnoticed at the time), received a brilliant experimental confirmation, culminating in the discovery in 1983–1984 of the vector bosons W and Z, which had exactly the features predicted by the theory (in fact, Salam and Weinberg received the Nobel Prize in 1979 for their work, and the discovery of the W and Z bosons has also led to the Nobel Prize being awarded to two experimenters working at CERN—Carlo Rubbia and Simon van der Meer). This experimental success has stimulated theorists to consider even broader unifications.

The second reason is purely theoretical. It is tied to the achievement of the central role of the concept of symmetry in the understanding of physical interactions. This way, we can think of a large symmetry that is, in principle, capable of unifying all physical interactions.[2] The progress of the theory on the plane of mathematical formalism allows us to tackle this problem, which is a priori extraordinarily complex.

Of course, this large symmetry cannot be exact, because in the case of the currently available energies, it is experimentally shown that the four interactions are very different from each other. The symmetry must be broken—in particular, it must be broken spontaneously, as one says in the language of physics (which means that the fundamental *equations* possess this large symmetry, although the *solutions* of those equations do not possess this property). A comparison made by Abdus Salam[3] can help us visualize the concept of spontaneously broken symmetry: let us imagine a round table set for lunch. All the diners have a salad at their right and one at their left, but nobody knows whether to take the one or the other. It is enough that one of the diners decides to take the one at his left; all the others will then do likewise. The right/left symmetry is thus spontaneously broken. Therefore, an agent is needed in order to cause the breaking. This role is assumed by the new particles that must appear in theories of unification.

To try to understand the functioning of a unifying theory, let us take the example of Salam and Weinberg's theory, in which the electromagnetic interactions and the weak interactions are unified in a single interaction—the *electroweak interaction*. At first, there are four messengers of electroweak interaction, which all have a zero mass because of the symmetry imposed. A new particle called the Higgs particle (named after Peter Higgs at the University of Edinburgh, who in 1964 invented a mathematical mechanism for achieving a spontaneous symmetry breaking) helps three of these four messengers to acquire a substantial mass. Leon Lederman called it the "God particle,"[4] but of course it has nothing to do with God. Higgs's particle was nevertheless the main argument in building the most powerful accelerator in the world, the Large Hadron Collider (LHC), whose essential mission was to find this somewhat miraculous particle. Higgs's particle was discovered in 2012.

The three messengers are the bosons W^+, W^-, Z^0 (+, −, and 0 indicate the electric charge of these bosons). The fourth messenger keeps a zero mass and is identified with the photon. The original symmetry has been broken, and two different interactions result—the electromagnetic one and the weak one.

Physicists have proposed even more unified theories, the *grand unified theories*, which attempt to address simultaneously the strong, the electromagnetic, and the weak interactions. The simplest model is that proposed in 1977 by Howard Georgi and Sheldon Glashow. In this model, they are treated on an equal footing, with a wider symmetry. As the symmetry is larger than that of Salam and Weinberg, the number of messengers for this interaction is also higher: there are twenty-four. In a first breaking of symmetry, twelve of them will gain a huge mass, so huge that they will never be noticed (this mass is obtained if one multiplies the mass of the proton 15 times by 10, $M = \text{proton}_m \times 10^{15}$). The other twelve will maintain a zero mass and are identified with four messengers of the electroweak interaction and eight gluons of the strong interaction. In a second breaking of symmetry, three of the four messengers of the electroweak interactions acquire a mass, following the mechanism described.

The unification of strong, weak, and electromagnetic interactions happens at a fabulous energy level (10^{15} times higher than the energy corresponding to the proton mass). According to Heisenberg's relationship, this energy corresponds to an infinitesimal distance (10^{-29} cm): if the proton were as big as the sun, this unification scale would be that of a speck of dust.

The energy corresponding to unification will never be achieved in our accelerators, but it was reached in the early big bang. Therefore, grand unified theories are very important for understanding what happened in the early big bang. The universe was likely a fireball, where a hellish temperature prevailed. An undifferentiated energy animated a huge number of quarks, leptons, and messengers, described by a single interaction. The fireball potentially contained the entire universe. Then, through a continuous cooling, different interactions gradually occurred. Originally, when a perfect symmetry ruled the world, all the stirring particles in the seething fireball had a zero mass. Today, the only observable messenger with zero mass is the photon: somehow, there is nothing but light, as a sign of the initially perfect world.

Particle and cosmos are therefore closely linked. The loop is thus closed: *by understanding the infinitely small, we understand the infinitely big.*

The unifying energy of strong, weak, and electromagnetic interactions is close to what is called *Planck's mass* (10^{19} times the proton mass), representing, as Planck noticed in 1900, the natural scale of unification of quantum physics and gravitational theory. It is therefore tempting to attempt the unification of *all* known physical interactions. Some attempts at such unification exist, such as *supergravity* theories, but they are still in a very speculative state and are experiencing some difficulties in their formulation.[5] These theories are based on the idea of *supersymmetry*, which unites particles with integer spin and those with half-integer spin.

It is interesting to note that some unification theories appeal to a space whose number of dimensions is greater than that of the world we live in (or think we live in). In 1921, the Polish physicist Kaluza proposed a theory of the unification of electromagnetic interaction and gravitational interaction that draws on a space of *five dimensions* (one dimension of time and four dimensions of space). Does this fourth dimension have any reality (see the following chapter)? It must be hidden somewhere, in one way or another. In 1926, the Swedish physicist Oscar Klein noted that, in order not to contradict the experimental data, this fourth dimension should wrap itself in a tiny size (10^{-32} cm), much smaller than the size of the atomic nucleus.[6] It is as if it did not exist! It is obviously not possible to visualize an extra dimension of space, because our sense organs are constructed according to a three-dimensional reality. The additional dimension of space has a rather mathematical reality. In any event, the unification fashion revived Kaluza and Klein's theory. For example, unified theories can be formulated in a space with at least *eleven dimensions* (one dimension of time and ten dimensions of space). In a sense, the corresponding symmetries are associated with *seven extra dimensions of space*. These seven dimensions have probably wrapped around themselves at 10^{-43} seconds after the big bang,[7] in a region of the order of 10^{-33} cm.

THE FINAL THEORY: SUPERSTRINGS?

Physics is a young science: it only appears as a distinct discipline around 1500, with the Renaissance. However, several centuries later, at the beginning of the third millennium, it claims to acquire the status of *A Theory of Everything*. Beyond the terminological ambiguity related to the word *everything*, the ambition is clear: to build a simple theory, *without free parameters*, that would represent, in combination with mathematics, the conceptual foundation of the pyramid of all scientific disciplines. The key challenge of this *final theory* is the unification of the general relativity theory and the quantum mechanics theory, a true Holy Grail for physicists.[8] All stairways of the physical universe would thus be related to each other, from the infinitely big to the infinitely small and to all known physical interactions—the strong, the electromagnetic, the weak, and the gravitational. They would be unified, like four branches of one and the same tree.[9]

The unification of physical interactions is, undoubtedly, the great intellectual adventure of modern science. Today, its impact on our culture is difficult to assess, but the idea that it would essentially influence the twenty-first century worldview is plausible.

THE UNIFICATION OF HEAVEN AND EARTH

What draws attention to the logical view of unification, a process that took three centuries to clarify, is the almost complete parallelism between two paths of unification: on the one hand, the weak and strong electromagnetic interactions, and on the other hand, gravity.[10]

The unification of terrestrial gravity and celestial gravity is the first major step, made thanks to Newton's genius. *Newtonian gravity* was the first theory of unification to connect, qualitatively and mathematically, the sky and the earth.

Two important events allowed Newton's theory to be surpassed by unification between space-time and matter: the idea of the *symmetry of calibration* and the *theory of restrained relativity*. Calibration symmetry is intrinsically linked to the invariance induced by conservation laws (as, for example, energy conservation). Einstein's restrained relativity theory, in turn, reconciles two fundamental physical facts: (1) the point of view of experimental observation and the formulation of physical laws of all moving reference systems are equivalent, with a constant speed in relation to one another, and (2) the speed of light in vacuum is constant (independent of the movement of the source). The consideration of the more general case of moving reference systems with a variable speed relative to one another becomes accessible. Thus, Einstein's *general relativity theory* is born, allowing the elimination of the separation between space-time and the moving body under the influence of the force of gravity: *mass and energy determine the curvature and geometry of space-time, and space-time determines the movements of material bodies.* In other words, matter and space-time are interrelated: they cannot exist independently. General relativity requires the existence of gravitational waves that have not yet been detected.

The next step in unification, via gravity, is quantum gravity. Indeed, general relativity is still in the realm of classical physics. It is now all about the unification of general relativity and quantum mechanics, two theories located in two radically different conceptual areas: classical physics, having a structural, nonprobabilistic aspect, and quantum physics, with a structural probabilistic aspect. Can two contradictory structures be reconciled? Several attempts at unification exist, but none have yet had any success. The major obstacle is the understanding of space and time, which play extremely different roles in the different theories proposed. The mathematical framework becomes increasingly more complex, increasingly more refined, and more obscure with regard to experimental design. There is still a convergence point between different models of quantum gravity: the manifestation of a new scale, called Planck's scale, where the fundamental units of length, time, and energy are defined in terms of gravitational constant G and of Planck's constant. On this scale, quantum gravitational effects are strong, and a theory of quantum gravity is inevitable. The space-time structure in particular is radically changed on Planck's scale. But nobody knows yet how this change occurs, and the most active researchers in the field must be content with a "semiclassical" approach in a very different framework than that of Planck's scale, where space-time can be found in its role of a passive receptacle of motion, which is, of course, an unsatisfactory theory.

In this way, a new step was taken by formulating *supergravity* theories.

Supergravity results from the union of quantum gravity and a new concept of unification—*supersymmetry*. A theoretical discovery of the 1970s, supersymmetry connects two fundamentally different classes of particles: *fermions*, particles of half-integer spin (1/2, 3/2, 5/2 . . .), and *bosons*, particles of integer spin (0, 1, 2,). (Let us remember that the spin is the intrinsic angular moment of the particle—from the English *spin* = to spin. According to the classical picture, spin describes how the particle rotates around itself.) Each known particle is assigned a new particle—its superpartner. These superpartners have not yet been discovered.

The relationship between supersymmetry and gravity comes from the fact that transformations that alter the spin of a particle simultaneously move the particle in space and time. The invariance related to local supersymmetry implies invariance in ratio to the movements in space-time (*the transformations of Poincaré*): the gravitational field is precisely designed to remove the distortions arising from Poincare's transformations. This is why local supersymmetry is actually supergravity. With the quantum gravity *graviton*—a messenger of the gravitational field, with a mass of zero and a spin of 2—superpartners were associated: fermions, with a mass of zero and a spin of 2/3. Due to the presence of the graviton, a mathematical miracle occurs: some infinite quantities, induced by the graviton, are annihilated by the infinite quantities induced by the gravitino. A first step toward a quantum and renormalizable field theory of gravity is thus made.

Supergravity is the last link in a long road to the unification of gravity that has Newton's gravity as a starting point.

CAN EVERYTHING BE UNIFIED?

Supergravity is rather a project than an achievement.

With regard to quantum gravity, we are also in the project phase. Grand unified theories represent the end of the logical chain of unification, whose starting point is electromagnetism and which passes through the Standard Model of strong interactions.

Despite its structural simplicity and high explanatory capacity regarding experimental design, the Standard Model is only a model and not a theory: there are a large number of free parameters, unconstrained by theory. "We do not know why particles exist, why they have a given mass, or why they are subjected to certain forces," writes Sheldon Glashow. "Our Standard Model is honest: it tells us that, in this context, there is no answer."[11] In turn, grand unified theories are convincing neither in theory nor in experiments. The simplest theories indicate, in fact, the existence of an *experimental desert*: nothing new will happen until the emergence of some huge energy, inaccessible to current and future accelerators. Murray Gell-Mann rightly observes that the term *grand unification* is not justified at all: "the theory contains more arbitrary constants than the Standard Model, in principle, always incalculable. Finally, gravity is not always given and it is as difficult as before to incorporate it."[12]

Despite these difficulties, progress continues, with the hope that for every complicated problem, there is a solution that is simple, thanks to the conceptual power of unification. Therefore, it is about the unification of the two extreme ends of the two parallel chains that we have described: supergravity and the grand unified theory, two somewhat virtual theories whose existence is dictated by the logical consistency of unification.

The first temptation will be to construct a grand unified supersymmetric theory similar in structure to a grand unified theory. In this field there is keen activity as well, but positive results remain to be seen: the difficulties each theory has in attempting to unify the fields are amplified when one attempts to unify the two theories.

An interesting attempt to unify the theories described so far is that of Laurent Nottale.[13] The relationship between the theory of relativity and quantum mechanics is as obscure as the relationship between spirit and matter. Nottale shows that the discovery of the laws of the world's unity represents rather an asymptotical project that will never be fully achieved. All we can do is to continually expand the truth area through more-unified theory. The new theory of Nottale (so original!) was dedicated to the subject with an openness that did not aim either to establish bridges between the two competing theories or to obtain classical physics from quantum physics. In fact, Nottale wanted to formulate a theory that was neither classical physics nor quantum physics but a *third* theory, whose specific cases, using certain scales, were the two theories. The technical means of achievement was the happy combination of mathematical fractals and the principle of relativity. Nottale noticed the shaky status of space-time in the quantum theory. Why artificially keep classical space-time in the quantum theory? Thus Nottale got to introduce a fractal space-time. He abandoned the sacrosanct dogma of differentiability, but its *scale relativity* succeeded the dialectical feat of strength in dealing with differentiability with the help of differential equations. On the technical level, one of the major innovations of Nottale's theory was the inclusion of the powers of spatial and temporal resolution at different scales in the very definition of coordinate systems. It is premature to say whether this theory will succeed or not. Very modestly, Nottale himself acknowledged that several centuries might be necessary before his theory was proved.

In the meantime, the only theory that seems to have a chance in the not too distant future to be a really superunified theory is *superstring theory*.

EVERYTHING IS VIBRATION

Superstring theory has two critical features. It does not cause *divergences*, which are infinite quantities present in most of the quantum field theories. Moreover, it does not show *abnormalities*, that is, inconsistencies in terms of quantum mechanics. These two features show that superstring theory is potentially able to unify general relativity and quantum physics.

The word *superstring* shows that, in this theory, elementary particles are not punctual particles but vibrating strings. These strings live in a world with one dimension: they have a length of the order of Planck's length (1.62×10^{-33} cm) but do not possess a thickness. In fact, *all particles in nature appear as vibratory modes of one and the same string*, which makes an enormous simplification to the complexity of the physical world and justifies the hope that *all* the parameters of the theory (dimensionless numbers) will be fixed by the theory itself, thanks to a universal autoconsistency.

Strings are open or closed (tiny loops). The internal symmetries of calibration are associated with open strings that have preserved charges at their two ends. When a string moves in space-time, it generates a two dimensional surface that records the entire history of that particle. This two-dimensional world is radically different, on the mathematical plane, from the one-dimensional world that characterizes the movement of the punctual particles present in other theories.

Consistency with quantum mechanics is gained by increasing the number of dimensions of our common space-time: there are twenty-six dimensions of space-time for bosonic strings and ten dimensions of space-time for fermionic strings.

There are only five viable superstring theories that are worth comparing with the great variety of grand unified theories. Of these five theories, that of *heterotic* strings has particularly drawn the attention of physicists. It is associated exclusively with closed strings. The theory incorporates supersymmetry in an aesthetic manner: for the bosonic area, ten dimensions correspond to the space-time dimensions; the other sixteen dimensions appear as unnoticeable internal degrees of freedom, which implies the existence of a new local symmetry of calibration.

Topology, a branch of mathematics, was born in the early twentieth century and plays a crucial role in superstring theory.

Topology is basically the science of *shapes*, the science of globalism. There is a big difference between geometry and topology: where geometry sees a multiplicity of shapes, topology can see only a single form. In particular, no distinction is made between shapes that can be continuously deformed into one another. Thus, in topology, one cannot distinguish between an oval and a circle or between a sphere and an ellipsoid. A sphere cannot be deformed in a continuous way into a torus—in this case, these are two different forms. Therefore, a particular form is characterized by a certain topological *invariance*, which it is possible to express by a particular numerical value: two different *shapes* will be associated with two different *numbers*.

The intuitive notions of *continuity* and *limit* get a precise, strict meaning in topology, thus associating continuity and discontinuity, discrete aspects and continuous aspects in the same formalism. Therefore, qualitatively, there is a great potentiality in topology for describing the world of particles, characterized precisely by the unity of the continuous-discontinuous pairing of contradictories. In particular, it is natural to think that the quantum numbers of particles (discrete numbers) may have a topological meaning.

When summing up all possible histories in superstring theory (in other words, when summing up all the two-dimensional surfaces associated with a particle), various surfaces appear as quantum fluctuations of a single *world-surface*. The sum of all histories involves an interaction between the superstring and the time-space. Here lies the conceptual opportunity to include general relativity in the theory as well.

The core problem is *the nature of space-time*. Quantum fluctuations that are extremely powerful deal with that problem: thanks to Heisenberg's uncertainty relations, space is no longer smooth, as in Einstein's general relativity. John Wheeler came up with the term *quantum foam* to describe the turbulence of quantum fluctuations. This is where we can find the theory of infinities in the theories that are still attempting today to unify gravity and quantum mechanics. From the start, superstring physics required a new geometry, beyond Riemann's geometry, which is the foundation of Einstein's theory and which upholds the concept of *distance between points*. The new physics, which no longer involves punctual particles, requires the formulation of a new mathematics. This new discipline is called *quantum geometry*.

The history of superstring theory is surprising, because its origin can be found in a theory that is completely outside unification theories: the *bootstrap* theory (see chapter 7). "Superstring theory resulted from a concept called the principle of the bootstrap," Murray Gell-Mann writes, "a nice and simple principle of autocoherence."[14] Physicists like me who have lived through the period of 1960–1985 are aware of this fact, but new generations of physicists completely ignore it.

The disappearance of the bootstrap from the physics scene coincides paradoxically with the attainment of a significant result that would constitute the basis of superstring theory: the use in 1968 by Gabriele Veneziano of a simple formula, invented long before Euler, describing, in a compact, unified manner, several major features of the hadron world. A fundamental property of duality characterizes this approach: the changes on one reaction path are related to the changes on another reaction path. I myself did my doctoral dissertation in this field of physics. Two years after Veneziano's discovery, Yoichiro Nambu, Holger Nielsen, and Leonard Susskind showed that Veneziano's formula can be interpreted in terms of the interaction between very low, one-dimensional vibrating *strings*. In 1971, Pierre Ramond, André Neveu, and John Schwarz elaborated an extension of the theory that included fermions: this approach is the recognized predecessor of superstring theory. Finally, in 1974, a crucial step was made: John Schwarz and Joël Scherk revealed that the theory in question predicted the existence of the graviton and, therefore, general relativity. These two researchers had the genius to turn a defeat into victory: a particle was raising serious questions about the validity of the whole approach, but Schwarz and Scherk were able to recognize here the graviton. Unfortunately, my friend Joël Scherk (1946–1980) died at a young age.

The course from Veneziano to Scherk opened the path toward the unification of all physical interactions through superstring theory. This was truly achieved, at least theoretically speaking, in 1984–1985, when Michael Green (from Queen Mary's College, London) and John Schwarz (from the California Institute of Technology) showed that the superstring theory with ten dimensions has no anomalies in terms of quantum mechanics. Physicists proudly call this *the first superstring revolution*, but until then, Green and Schwarz had been working in complete solitude on the outskirts of the activity of most quantum physicists, who, dizzy with the successes of the quantum theory of fields, were giving no chance to the new theory. Today, in the field of particle physics,[15] this theory directs most young researchers' work.

Superstring theory does not tolerate any arbitrary parameter: all pure numbers must be deduced from the theory itself. The only parameter accepted is, of course, the one that must determine our own measurements, a scale parameter, namely the *tension* of the superstring. Scherk and Schwarz have observed, since 1974, that this tension must be enormous, with no possibility of comparison with the tension of our terrestrial strings, as, for instance, those of a hammock. The relationship with gravity implies that this tension is that of Planck: 10^{39} tons. The mass scale is set by Planck's mass: 10^{19} times the mass of the proton, which already corresponds to the mass of a macrophysical object, such as a grain of pollen. The size of a string is fixed at a very small but *finite* length, Planck's length: 10^{-33} cm. It is this finite size that allows superstrings to be insensitive to quantum fluctuations. But *in order to observe those superstrings, at such small distances, an accelerator the size of the universe would be required.* The corresponding energy is not enormous—it is of the order of the energy consumed by a household—but the problem consists in the focus of this energy on such a small size, infinitely smaller than that of household appliances. Will humanity ever reach such a stage of technological development?

Superstrings radically alter our conception of physical reality. The superstring, the fundamental entity of the new theory, represents a spread object in space. Therefore, it is logically impossible to define *where* and *when* superstrings interact. This feature is entirely in the spirit of quantum mechanics. On the other hand, their finite size implies the existence of a *limit* to our ability to probe the reality. Our anthropomorphic convention of distance no longer applies. Neither the universe nor any of its objects have any sense beyond this limit. Finally, the spatial dimensions are of two kinds: large, vast, visible (like the three dimensions of what we consider as our space) and small, *wrapped around* themselves, invisible. This totally changes our vision of nature in relation to our ordinary perception.

On a purely theoretical level, some extra dimensions of time could also exist. Some theoretical works have explored this extremely disturbing possibility, which would mean that superstrings could *travel in the past*. Such a possibility is so far from our human experience of time that physicists prefer to think that all of the extra dimensions are spatial dimensions.

The supplementary dimensions of space have numerous consequences. The masses and the electric charges of the particles, which we measure in our three-dimensional space, are actually caused by the extra dimensions. Superstrings can wrap around a spatial dimension. Wrapped superstrings and unwrapped superstrings have different properties: each class of strings has its role in building the world as we know it. There is, of course, a vibratory energy, but there is also a *winding energy*, which depends on the radius of the dimension in question but also on the number of wrapping shifts. The unwrapped strings can generate particles of mass zero, such as the photon, but the wrapped strings are limited by the non-zero mass.

A paradoxical property of *duality* reigns in the world of superstrings: physics in a certain range (a circle of a certain radius) is strictly equivalent to physics in a circle with a radius that is the inverse of the precedent. A circle with a radius of one meter is equivalent to a circle with a radius of one centimeter. What if this happened in our universe? If the three dimensions of our space, instead of continuing indefinitely, wrapped around a huge circle, our universe, rather than expanding, might well shrink! "If we accept light string modes," says Brian Greene, "the universe would be vast and it would expand; if we accept heavy string modes, it would be small and it would shrink. And there is no contradiction: we have two definitions, distinct but equally valid, of the distance. The first of the two definitions is more familiar to us because of our technological limitation, but each one is an equally acceptable concept."[16]

Superstring theory has fanatic supporters but also fanatic detractors. No one denies the aesthetic beauty of the theory. What is disputed is its predictive power. It is not known what the fundamental equations of this theory are. The mathematics used in superstring theory are so complex that even mathematicians are often left confused. "So far, they have not made the smallest experimental prediction," says Sheldon Glashow, Nobel laureate in physics. "In the meantime, the historical connection between experimental physics and theory has been broken. . . . Are reflections on strings really more suited to mathematics departments or even to divination schools rather than to physics departments?"[17]

Why would it matter? Progress continues, because physicists cannot tolerate the multiplicity, however small, of superstring theories. Edward Witten, emblematic figure of the beginning of twenty-first century physics and holder of the Fields medal (equivalent to the Nobel Prize in mathematics), speaks of a *second superstring revolution*, which according to him took place in 1994–1995.[14]

THE MYSTERY THEORISTS

The new idea is that of a generalized *duality* that manages to tie together six fashionable theories: the five superstring theories and supergravity with eleven dimensions. These six theories appear as limited cases of one and the same theory that alone would deserve the status of a "Theory of Everything." The theory also has a name: the M theory (the name invented by Witten); the letter M signifying, according to different authors, *Magic*, *Mystery*, *Mother*, *Matrix*, or *Membrane*. Of course, only the last two words are technical terms. A psychoanalyst would find a good subject of study in an analysis of the strange ways of scientific terminology.

Witten's great discovery is the *eleventh* dimension of space-time: the M theory requires ten dimensions of space, not nine, as in superstring theories. It is precisely this theory that allows superunification. The new dimension does not characterize the vibration of strings but their structure. The five previous theories saw but ten dimensions, because for their conceptual framework, the eleventh dimension was too small, invisible. Thus, strings are no longer one dimensional, but two dimensional—*membranes*.

The situation is actually both more complex and simpler. A true *dimensional democracy*, a modern version of the bootstrap's nuclear democracy (see chapter 7), reigns in the magical universe of the M theory. Strictly speaking, there is no other fundamental entity than the dimensional superstring. All kinds of mathematical-physical objects populate the M universe: points (zero-branes), one-dimensional strings (uni-branes), membranes (duo-branes), three-dimensional bubbles (tri-branes), and so on, up to ninebranes. The dimensional democracy is a democracy of *branes* (according to Paul Townsend's phrase). It has created the impression of finding oneself back in Flatland, the world imagined by Edwin Abbott.[19]

A subtle duality allows the interlinking of the five superstring theories. The strong coupling of one of the five theories corresponds to a weak coupling of another theory from the same family (the "coupling" is related to the likelihood of splitting a string into two strings).

The zero-branes have an interesting status, because for them there is no space, no time, yet nevertheless, they could constitute *the genetic code of space-time*. This draws on the *noncommutative geometry* developed by Alain Connes.[20]

Nowadays, it is impossible to predict what the future of the M theory will be, because this theory is, mathematically speaking, extremely complex. In any case, a new deep idea appears within this path of unification: quantum mechanics will make possible a new symmetry of the natural world. In Edward Witten's opinion, this symmetry will mark a new stage of physics.

SEEKERS OF TRUTH

Two conclusions seem to loom in the light of the various attempts at unification.

The first conclusion is collective, shared by most members of the community of particle physics: quantum mechanics appears to have a particular and unusual status among physical theories. It is more than a theory: *quantum mechanics is a general framework that any physical theory must be part of.*

A second conclusion is less shared by physicists: fundamental physics seems to have changed its nature, speaking in terms of a discipline. It is more oriented toward an area of interface between mathematics and metaphysics.

The philosophical and theological implications of superstring theory are tremendous if this theory is right. The universe was already very rich, before the formulation of this theory: 10^{11} galaxies and 10^{11} planets in each of these galaxies! But the universe shifted, in string theory, toward a *megaverse* or *multiverse* (a multiple universe) full of an incredible number of *pocket universes*, our own universe being just one of these pocket universes.[21] String theorists consider uniqueness to be just a myth,[22] but it was precisely this uniqueness that was the starting motivation of string theory: physicists thought that string theory would prove to be the *only possible theory of everything*. However, the net result is just the opposite: *there are more than 10^{500} (one followed by five hundred zeros!) possible theories*, that is, more than 10^{500} worlds governed by different interactions and physical constants and populated by different particles. The fundamental reason for this proliferation is the existence of extra dimensions: they curl up in a huge number of possible ways. This huge number of choices for the geometry of extra dimensions leads to a huge number of possible free physical constants.[23] The *landscape* of string theory looks like a science-fiction landscape. Under the pressure of these mathematical results, string theorists converted a failure into a triumph. The strange fact is that the nature of scientific prediction radically changes in the framework of string theory. Because of the presence of a huge number of pocket universes, any given experimental data will find an explanation in one of these universes. More clearly, it means that all power of prediction is lost!

The theological implications are also striking. It is highly probable, given the huge number of possible universes, that there are other planets than the Earth populated by humans. As Lee Smolin points out, recalling the case of Giordano Bruno (1548–1600): "If there are other planets with other people on them, then either Jesus came to all of them, in which case his coming to Man was not a unique event, or all those people lose the possibility of salvation! No wonder that the Catholic Church burned Bruno alive."[24]

Maybe today we are about to witness the birth of a *metaphysical mathematics*, together with the birth of the M theory, which is the precursor, in the long term, of a *mathematical metaphysics*. Despite their conservative spirit, physicists like revolutions: "Science does not relate to the status quo, but to a revolution," says Leon Lederman, Nobel laureate in physics. "When a revolution occurs, it extends the domain of validity of science, but, at the same time, it can have a profound influence on our vision of the world."[25]

The entire problem is that the string revolution remains unachieved after long years of research. Lee Smolin wrote an entire book aiming to "demystify the claims of string theory."[26]

From a philosophical point of view, it is natural for physics to face its own methodological limitations. These were well analyzed by John Barrow.[27] In particular, it would be surprising if Gödel's theorem, about which physicists speak so little, did not have any impact on scientific constructions. In fact, the mathematics used in physics includes arithmetic, and thus, logically, physical theories should be subject to the findings of Gödel's theorems[28]: *a sufficiently rich system of axioms without internal contradiction is necessarily open (there will always be true results which cannot be proven), and therefore a rich enough closed system is necessarily contradictory.* Perhaps this is the source of the difficulties encountered by unification theories, including superstring theory, in the area of experimental prediction.[29]

"Maybe we have to accept, after having reached the most thorough understanding science can provide, that certain aspects of the universe are still unexplained," writes Brian Greene, who is far from being a mystic (another word beginning with M). "Maybe we will have to accept that some of its features are due to a juncture, to chance, or even to some divine work . . . ; if we reach the absolute limits of scientific explanation—which would be neither a technical barrier, nor a current frontier in the progress of human knowledge—this would be a unique event, for which past experience has not prepared us." And Greene concludes: "We are all, each in our own way, seekers of truth."[30]

The road toward unification in physics seems to be an endless route, but it marks a prolific intellectual adventure, of vital significance for the understanding of the nature of Nature and of the nature of knowledge.

CHAPTER SIX

THE STRANGE FOURTH DIMENSION

Was Saint Augustine truly wrong when he asserted that we are not able to define, and therefore represent, time? Physicists, from Galileo onwards, have argued that time may be thought of mathematically. Einstein's genius led him to infer that time is a *dimension*, like the three dimensions of space, in a larger space with four dimensions.

The contradiction between these two views is only an appearance. In fact, no one, not even mathematicians or physicists, can represent the *reality* of a space with several dimensions. Try the simple exercise of adding a fourth dimension to the familiar three dimensions—length, width, and height—of the room you are in now, and you will find that it is impossible to *see* this new dimension. Our brain has been shaped by the constraints of natural selection: buffalos and lions were killed in three dimensions and not four! But not being able to see an extra dimension does not mean that it does not exist. Are we, perhaps, led by the invisible in our visible world?

The extraordinary adventure of the extra dimensions of space began with a German mathematician, Georg Bernhard Riemann (1826–1866). In 1854, when he was twenty-eight years old, he presented a thesis, *Über die Hypothesen, welche der Geometrie zu Grunde liegen*, which has influenced and continues to influence not only mathematics but also physics. In this thesis, which was only published after his death in 1868, he introduced the concept of a „differentiable manifold with n dimensions," which extends our ordinary conception about space. In the same thesis, Riemann launched another revolutionary idea, long before Einstein: the interaction between space and matter. Riemann's thinking is actually guided by the principle that *the laws of nature become simpler and unified when they are considered in a space with more dimensions than our usual space*. Today, we find exactly the same idea in advanced physics—that of superstrings (see the previous chapter).

Having a fragile mind—he suffered from chronic depression, which tormented him terribly—Riemann died at the age of forty after having left the chair at the University of Gottingen in 1862.[1] No one can say whether depression is what opened the channels of his sensitivity to extra dimensions or if it was the contemplation of these dimensions that caused his mental disorders. But the truth is that these dimensions have something troubling.

Oddly, physicists did not pay attention to Riemann's ideas, and physics followed a parallel, slower path. Instead, artists, writers, and mystics of all kinds literally pounced on the idea of extra dimensions. Lewis Carroll (the pseudonym of Charles Lutwidge Dodgson [1832–1898], mathematician, logician, photographer, novelist, essayist, deacon of the Anglican Church, and professor at the University of Oxford) actually draws directly upon Riemann for *Alice in Wonderland*: how could Alice walk through the mirror if there were no extra dimensions of space?

It is hard to imagine today, a century later, this incredible vogue for multidimensional spaces that lasted for more twenty years around 1900.

Thus, in 1884, the theologian Edwin Abbott (1838–1926) wrote a novel, *Flatland: A Romance of Many Dimensions*,[2] which became an immediate best seller precisely because of the public passion for multidimensional worlds. In his novel, Abbott described the adventures of a two-dimensional being that is pulled out of his world by a three-dimensional being. Upon returning home, the two-dimensional being testifies to the existence of other worlds, of other realities, at the risk of his life (because great priests used to murder all embarrassing witnesses). The allusions to our own world are transparent. It is obvious that two-dimensional beings are unable to do certain things that three-dimensional beings can do without any problem. For example, the beings from Flatland cannot have a digestive system! Imagine a creature that can live in two dimensions and only in two dimensions, like a sheet of paper without thickness. A digestive tube may dislodge it completely into two parts that are unconnected to one another. In the same way, we three-dimensional human beings are, perhaps, in the same situation as an inhabitant of Flatland: other worlds and other entities are present here without us even suspecting their existence. However, they are here, for better or for worse.

Another important example is the famous book *The Time Machine*, by H. G. Wells (1866–1946), published in 1894, which describes, through a machine that can make us travel in time, the England of the year 802,701. Beyond the fact that, through this text, Wells marked the history of science fiction, he has the merit of having conceived, before Einstein, of the fourth dimension as time, even though most fans of the fourth dimension had conceived of it as just another spatial dimension. It is remarkable that the expression of Wells, the *time machine*, was taken up by today's quantum physicists in a direction of research that belongs to science and not to science fiction. It was only in 1988, a century after Wells, that the first scientific proposal of a *time-generating machine* was published in the prestigious international journal *Physical Review Letters*.[3] We will presently see what is meant in science by "time-generating machines" that involve travel in time. However, to avoid the possible disappointment of the reader, I must say that this trip will not be at all comfortable. We will not be able to sit in front of the computer, press a few keys, manipulate some buttons, flash a few light bulbs, and see on the screen what will happen in 2,000,001 while remaining inside the warm comfort of our office. According to our physicists, people belonging to a highly developed terrestrial civilization will be able to build a time-generating machine based on Einstein's *current* theory of general relativity: they will place inside each of two different boxes a pair of very wide boards, between which they will establish a strong electric field. This field will interact with space-time and thus alter its structure. According to Einstein's theory, a *hole* is thus created in the space connecting the two boxes, a hole of space-time (a *wormhole*, as today's physicists call it, with a rather earthy imagination). Then, our friends will place one of the boxes in a rocket, while keeping the other one on Earth. Thus, the hole links two regions of space where time is different: time runs slower in the first box. A person who reaches one end of the wormhole will be instantly drawn to the other end of the wormhole, where time is necessarily different. Thus, the person will travel into the past or into the future.

The climax of the vogue of the fourth dimension is marked by writings of an extravagant character, Charles Howard Hinton (1853–1907), who was both a mathematician and a philosopher. Hinton started with an enviable academic career: he studied at Oxford and became a professor at Uppingham School. He began to be passionate about the fourth dimension issue, and he married the widow of George Boole (the founder of Boolean algebra, which is so common in our computers today). God knows what pushed him to marry a second woman as well—Maude Weldon, with whom he spent several days in a hotel and with whom he had twins. After having been convicted of bigamy in 1886, he was fired from his job and imprisoned for three days. Then Hinton and his first family escaped to Japan. In 1893, he could be found inside the mathematics department at Princeton University, where he invented a baseball pitching machine that is still used today. Once fired, he ended his life in a patent office (like Einstein!). He died suddenly at the age of fifty-four while giving a toast to the honor of "philosophical women" at the banquet of a philanthropic society in Washington.[4]

Hinton's article "What is the fourth dimension?" appeared in 1880 in a magazine of the University of Dublin and had a resounding success. His writings were collected in 1904 in the book *The Fourth Dimension*,[5] a stunning effort in writing due to the constant back and forth between mathematics and metaphysics.

As we read this book, we can understand the origin of Hinton's passion for spaces with n dimensions. It is simply all about human dignity.

Of course, like everyone else, Hinton knew very well that it is impossible to directly see these strange spaces. But let us suppose that the fourth dimension truly exists. Then perhaps we are four-dimensional beings ourselves, but, for one reason or another (natural selection?), we became blind to the fourth dimension, just as those blind from birth cannot see or imagine colors. Thus everything happens as if we were the owners of a sumptuous castle but lived our entire lives in a squalid apartment, completely ignoring the existence of our own palace.

There is also another possibility: we are really three-dimensional beings. In this case, we are puppets in the hands of four-dimensional entities, our gods (or our demons?), just as we are gods (or demons?) for the hypothetical beings in Flatland. Our dreams, our loves, our hopes—are they, perhaps, but smoke, simple thoughts in the mind of our multidimensional gods? Is the universe itself a dream of these gods?

We see very well, both in the field of mathematics and in the field of metaphysics, what our human dignity consists in. Is not it strange to find Hinton's considerations, formulated in the early twentieth century, in our imaginary of the twenty-first century, as expressed in a cult film like *The Matrix*? In our computer era, will the multidimensional god be a cosmic computer that will consider us virtual slaves and want to punish the last opponents—those who know the reality of things and fight for a possible release?

Hinton expresses our human position in admirable phrases: "May we not compare ourselves to those Egyptian priests who, worshipping a veiled divinity, laid on her and wrapped her about with ever-richer garments and decked her with ever-fairer raiment. So we wrap 'round space our garments of magnitude and vesture of many dimensions. Till suddenly, to us, as we to them, as with a forward tilt of the shoulders the divinity moves, and the raiment and robes fall to the ground, leaving the divinity herself revealed, but invisible; not seen, but somehow felt to be there."[6] Hinton states the solution for our release, which is both mathematical and physical: "The true apprehension and worship of space lie in the grasp of varied details of shape and form, all of which, in their exactness and precision, pass into the one great apprehension."[7]

This is the key to the methods and to the countless exercises invented by Hinton in order to cultivate our brain, in order to get it used to entering the lassoes of the four-dimensional world: the visualization and the perception of all the *details* of shapes and configurations. Thus, for example, an inhabitant of Flatland cannot perceive a cube, but he or she can see and understand its projection on a sheet of paper. One side of the cube rests on the sheet of paper. If one cuts the cube along its other lines, the result is a cross composed of six squares in the bidimensional space. The inhabitant of Flatland cannot rebuild the cube, but we can do it. Then the Flatlander will suddenly see five squares from his or her world disappearing and will believe that this is a miracle. Similarly, we cannot see a hypercube—that is a four-dimensional cube. But we can see and understand very well its projection in our space, which consists of a cross made of eight superimposed cubes. This cross has a name—*tesseract*—invented by Hinton and entered into the common English language. Hinton's four-dimensional cross fascinated Salvador Dali, who, inspired by it, painted his famous *Christus Hypercubus*. We see in this painting Christ (Dali?) crucified on this hypercube cross, levitating above the floor, itself made of squares. At his feet, Mary (Gaia?) prays. What highest praise brought to the fourth dimension!

Hinton's projection methods are still used by theoretical physicists to represent spaces with several dimensions, even if they completely ignore the name of Hinton.

He even built a kind of three-dimensional retina of a four-dimensional being: an arrangement of 46,656 cubes in a huge cube, thus made of 36 × 36 × 36 cubes, each with sides of 3 inches. This structure has actually only 216 distinct components. Thus, Hinton was able to visualize the invisible.

He also had a brilliant idea regarding the fact that we cannot directly see the multidimensional world. If an extra dimension is invisible to us, this means that *it is wrapped in an infinitesimal region of our space*. This exact idea—called "compactification"—can be found in current grand unification theories (see chapter 5) that ambitiously aim to bring together the four known physical interactions! In this case, light would be the vestige of these extra dimensions, which explains its special role in Einstein's relativity. Hinton goes even further, inferring that if an extra dimension is wrapped in an infinitesimal region, then the brain cells—our neurons—must have consistency in this extra dimension. We are truly four-dimensional beings whose brains are naturally and materially capable of perceiving the fourth dimension. This property could explain, after all, why mathematicians are able to play with multidimensional spaces even if, like ordinary mortals, the mathematician is unable to see them.

Hinton and the fourth dimension had popular success beyond anything that we could imagine today. In 1909, the very serious magazine *Scientific American* organized a contest featuring a prize of $500 for the best popular explanation of the fourth dimension. The editors received letters from all around the world. It is amusing to note that none of them mentioned Einstein!

Hinton's ideas rapidly migrated to Europe, and especially to Russia, due to the philosopher Pyotr Demianovitch Ouspensky (1878–1947). Unlike Hinton, Ouspensky focuses on the relationship between consciousness and the fourth dimension and identifies this new dimension, following Einstein, as one of time rather than space. For Ouspensky, the fourth dimension opens access to the mystery of our consciousness, thus allowing us to undergo real personal development comparable to the biological evolution that the human being has seen over time. Associated with the anxious and tireless spirit of Russia before the revolution, this explains Ouspensky's extraordinary influence in intellectual and spiritualist circles and especially among writers and painters. A prolific author, Ouspensky published in his essay *The Fourth Dimension* in 1910 in St. Petersburg, and it instantly spread throughout Russia. A year later, this study was used in his major work *Tertium Organum*,[8] of which the second edition appeared just before the Russian Revolution. In 1920, the English translation of his book ensured Ouspensky's strong international reputation in academic environments (in fact, his papers can be found at Yale University Library's Manuscripts and Archives Department). The writer J. B. Priestley (1894–1984) drew upon Ouspensky's ideas half a century after the appearance of *Tertium Organum* in his stunning book *Man and Time*,[9] which is illustrated with beautiful historical representations of time.

Almost all Russian intellectuals used to speak passionately about the fourth dimension, thanks to Ouspensky's writings. Dostoevsky drew upon them for his book *The Brothers Karamazov*: Ivan Karamazov refers to extra dimensions in a discussion about the existence of God.

Lenin himself got involved in the dispute. He was appalled by the idea of the fourth dimension, which appeared to be part of spritualism, which he saw as a major threat to the revolution that was being prepared. In the huge text *Materialism and Empirio-criticism*, which he wrote during his exile in Geneva, Lenin declared war on the fourth dimension, pivot of the mysticism and the metaphysics of that age. Lenin had nothing against the mathematicians who explored multidimensional spaces, but, he wrote, "The Emperor can only be overturned in three dimensions"! He proved this later, in the bloody revolution. But did he truly prove it? Did not the tsar take his revenge in 1989?

The most important influence of the fourth dimension was manifested in painting. In a well-documented and thorough study, Jean Clair demonstrates the crucial importance of Ouspensky in the suprematist painting of Kasimir Malevitch (1879–1935). Clair analyzes works such as *The Musical Instrument, Lamp, The Portrait of Matiushin, An Englishman in Moscow,* and his *Movement of Color Masses in the Fourth Dimension* series, as well as discussing the theory in his text.[10] The three major consequences in the field of representation are, according to Jean Clair, perspective denial, measurement denial, and denial of the object as a whole.

Furthermore, the art historian Linda Dalrymple Henderson[11] considered that the fourth dimension is a unifying theme of modern art. For cubists, for example, the exploration through art of multidimensional worlds is a reaction against the straps of the narrow, mechanistic, and positivist materialism of the nineteenth century. It is interesting that this insurrection occurred in the name of science.

For instance, let us have a look at the Picasso's famous painting *The Portrait of Dora Maar*. The duplicate face of the woman is exactly what a four-dimensional being would see when looking at a woman from our three-dimensional world, because he would be able to see this face from all angles at once. For Picasso, the fourth dimension is a space-like dimension, not a temporal one. Marcel Duchamp is the artist who drew near Einstein: his painting *Nude Descending a Staircase* shows a large number of slightly blurred images of the same woman, like successive flashes. If the fourth dimension is time, a four-dimensional being would see all the successive images *simultaneously*. The past and future have no distinct existence for him. In space-time, everything is there, always there.

Jean Clair relevantly notes that one and the same theme—the fourth dimension—gives rise to two distinct currents in painting. One of these currents—represented by Gleizes, Metzinger, Picasso, Villon, and Duchamp—uses this theme as an excuse for purely formal intellectual games. A second current—represented by Malevitch, Kupka, Lissitzky, Mondrian, and van Doesburg (and, I would add, Rothko)—takes this thesis in support of a spiritual quest, far from any formal game. When looking at one of Rothko's paintings, I am not assaulted by a more or less intelligent illustration of multidimensional spaces. My exit from the three-dimensional world operates through a subtle and invisible vibration that emanates from the painting and penetrates me, and, in turn, I get inside the painting: its vibrations and my being make one.

The issue of the fourth dimension is not limited, of course, to painting. It is also fully present in literature. Besides the authors already cited, we can mention Oscar Wilde, Marcel Proust, and Joseph Conrad. In music, it inspired composers such as Alexander Scriabin, Edgar Varese, and George Antheil. In fact, the entire culture has been influenced by this issue.

But let us get back to our starting point, to Einstein.

For half a century after the formulation of Riemann's revolutionary mathematical ideas, nothing remarkable happened in this context in physics. All this tumultuous debate around the fourth dimension in the world of art, literature, and philosophy proved to be completely sterile in terms of physics. Some cynical minds believed that the fourth dimension was good for spiritualists and mediums and for mathematical games but had no real existence. At this time, an obscure physicist named Einstein entered the stage.

Albert Einstein (1879–1955) demonstrated, beyond doubt, that this dimension is real, provided that we interpret it as a dimension of time associated with a very special geometry. Paradoxically, it was not Einstein who discovered this geometry but one of his teachers, the Russian Hermann Minkowski (1864–1909). It is amusing to note that Einstein, considered by everyone as the prototype of the modern theoretical physicist, actually had a very limited knowledge of mathematics. He certainly had fabulous and unrivaled insight, but he always had to find the mathematical formulas that would fit with his findings somewhere else. When Minkowski gave his memorable conference speech at Köln in 1908, presenting the geometry adapted to the restrained relativity theory,[12] Einstein was very skeptical about what seemed to him an unnecessary pedantry. But he soon realized the full significance of Minkowski's discovery.

Space-time hits the world's stage as a *block universe*, where there is no place or time but only *events*. Space and time are intertwined—they do not exist separately. Time and space are like shadows. Only their union gives meaning to what we may call an *independent reality*. In the space-time continuum, there are neither space-like landscapes nor temporal landscapes. This is a more radical view than that of Picasso or Duchamp. In this block universe, no one can really say "now," because "now" for one observer is not the same as "now" for another observer. Simultaneity does not exist. Our own lives certainly fill only a tiny portion of this space-time, but they explore what is *already there*. They are only "universe lines." In this universe, one cannot say "my friend died when he was fifty-seven years old," because this is a meaningless statement, since time is a mere illusion. Becoming does not exist. A strange vision, completely at odds with our intuition! But this is how the world works, if the physics viewpoint is right. Even if they are convinced that time is merely an illusion, this does not prevent physicists from behaving in everyday life as if the damned time did exist. Like other human beings, physicists are steaming with annoyance in traffic jams because they are "wasting time," or they are fighting without mercy in order to improve their careers and "gain time." But how could one gain or lose something that does not exist?

We cannot, of course, view this space-time continuum, but we can represent it mathematically by means of Minkowski's diagrams, which are, indeed, a variant of Hinton's diagrams. More specifically, time is designated by a vertical axis and space by two horizontals that are perpendicular on the time axis. The "now" of one observer will thus be shown by planes that are parallel with the plane that consists of the two horizontals, and the

THE STRANGE FOURTH DIMENSION

"now" of another observer will be shown by other planes that are parallel to each other but that are inclined in relationship with the plane of the two horizontals. Thus, one may "see" why there is no absolute "now" in Einstein's universe. And one can do a lot of other useful things with these diagrams of Hinton-Minkowski.

Nevertheless, the time dimension still plays a very special role in relation to the other three dimensions. Time is not fully spacelike. In order to understand this, we must consider the notion of *distance in space-time*. In our world, we know what the distance between two points is: we take a ruler and measure this distance. We also know what a time interval (or temporal distance) is: we look at the clock at two moments of time and we measure the interval. But in space-time, things are more complicated. Minkowski tells us that we must do the following operations in order to measure a distance. We take the time interval and multiply it by the speed of light; we thus get a common spacelike distance in which a second will be equivalent to a distance of 300,000 km, a rather large number, but we have no choice, because the speed of light is 300,000 miles per second. We calculate the square of the resulting number. Then we take the square of the spatial distance (in kilometers). We subtract the first number from the second number. Finally, we take the square root of the resulting number, and we have the distance between two events.

One can already see that time is associated with the negative sign ("–"), which is unusual for those who know the theorem of Pythagoras. But there is nothing arbitrary about this; it is related to a deep property of the theory. No matter how relative relativity theory is, it is still in search of *invariants*. The speed of light is the first invariant, at the foundation of the theory. The distance between two events, which has just been defined, is another invariant. A third example is the mass of a particle measured in the reference system at rest. There are also other invariants. These invariants are very useful in particle physics.

Let us try to apply the definition of distance to a few simple examples.

Let us consider the Eiffel Tower at 09:00:00 a.m., on December 25, 2001, and, on the same day, the center of the sun at 09:06:40 a.m. What is the distance in space-time between these two "events"? It is very easy to calculate. The distance in space between the Eiffel Tower and the center of the sun is 150 million km, and therefore its square is 22.500 billion km^2. The time interval of interest is 6 minutes 40 seconds, which is 400 seconds. If we multiply it by the speed of light, we obtain 120 million km, of which the square is 14.400 billion km^2. The difference between 22.500 billion and 14.400 billion is 8,100 billion. The square root of this last number is 90 million. So 90 million km separate the Eiffel Tower and the center of the sun in space-time. This distance is 60 million km smaller than the distance between them in our space! If we ask ourselves, "What is the distance in space-time between the Eiffel Tower at 09:00:00 a.m. and the center of the sun at 09:08:20 a. m?" we get a response that is even more surprising to our common intuition: this distance is strictly zero, even if the separation in space is 150 million km! How is this possible? In fact, the answer is very simple. Within 8 min 20 sec (that is, 500 seconds), the light travels 150 million kilometers, which corresponds to the distance in our space between the Eiffel Tower and the center of the sun: the distance in space and the distance in time between our two events are simply equal.

Things get a little bizarre if we ask the slightly perverse question: "What is the distance in space-time between the Eiffel Tower at 09:00:00 a.m. on December 25, 2001, and the same Eiffel Tower at 09:06:40 a.m. on the same day?" In this case, the distance in space is zero, so one has to extract the square root of a negative number. This means that the distance in space-time is expressed through a number called "imaginary." Here is something that totally frustrates our intuition, because we are accustomed, in our space, to either positive or zero distances but not to imaginary ones. Of course, the imaginary number is only called imaginary: in fact, it is as real as our well-known real numbers. It is important to briefly recall what these imaginary numbers are (their history covers several centuries—see chapter 15), because different people may have different ideas (right or wrong) in their minds when the phrase "imaginary number" is uttered or written.

It has been well-known for a long time that the square of any number (positive or negative) is positive. Thus, calculating the square root of a negative number seemed an absurd, impossible operation. However, in the sixteenth century, Gerolamo Cardano, one of the most fascinating characters in the history of science, dared to introduce these new numbers (the resulting square root extracted from a negative number). As noted by Jacques Hadamard in his *Essay on the Psychology of Invention in Mathematical Field*, these numbers seemed closer to madness than logic, but, paradoxically, they enlightened all of mathematics (see chapter 15).

Cardano, the great Italian Renaissance algebraist, is best known as the inventor of "Cardano's joint" (or "the Cardano"—a suspension system that provides an invariable position for a hanging body). He was a professor of mathematics at Milan and a professor of medicine at Bologna and Pavia. Married to the daughter of a bandit, he had two sons who became criminals. His life was an uninterrupted vagrancy. He was pursued and spied on by many governments, and in 1570 he was indicted for heresy and arrested. The man who called himself "the seventh doctor after the creation of the world" had a blind faith in the revelations of dreams and had suffered from hallucinations since childhood. In fact, he thoroughly described his dreams and hallucinations in one of his many books, *De Somniis*. The man who established the horoscope of Jesus Christ was at the same time one of the most important scientists of his time. And the man who believed, like Paré, in the existence of a „bird of paradise" (a bird without legs that feeds itself solely with air and dew) was also the author of *Liber de ludo aleae*, the first systematic presentation of probability calculus.

Therefore, it is not too surprising that Gerolamo Cardano had the courage to accept the existence of „imaginary" numbers. In 1545, in the book *Artis magnae sive de regulis algebraicis* (called *Ars Magna*), Cardano studied the solution to the third degree equation and introduced, in his famous formula, these numbers, so-called "impossible numbers." The term *imaginary numbers* was invented later, in 1637, by Descartes. Finally, the notation "i," the initial for the word *imaginary*, symbolizing the square root of (−1), was introduced in 1777 by Euler.

The reticence of mathematicians about imaginary numbers was considerable, affecting the constant effort to justify the "reality" of these numbers. The legitimacy of imaginary numbers was only acquired in 1806, when Robert Argand discovered a geometric representation of imaginary numbers that highlighted the similarity in nature between real numbers and imaginary numbers. He associated the concept of "direction" to that of a given category of numbers: if real numbers are represented as points on an axis, then imaginary numbers are points on the perpendicular axis. Thus, the mysterious "i" is only an operator of perpendicularity. The road from Einstein to Cardano and Planck is long, but modern physics is unthinkable in the absence of the imaginary numbers.

Let us resume our example: the space-time distance between the Eiffel Tower at 09:00:00 a.m. on December 25, 2001, and the same Eiffel Tower at 09:06:40 a.m. on the same day equals the imaginary number "i" that multiplies the distance traveled by light in 400 seconds: it is therefore of 120 million iKm. . . . There is no mystery here: these imaginary distances in space-time correspond to the case in which the temporal separation between events is *bigger* than their spatial separation.

It is now possible to understand a little better the origin of the "−" sign in Minkowski's definition of the distance in space-time. Indeed, the square of "i" is precisely −1. Therefore, the square of the distance in space-time can be expressed as a *sum* of squares, which reminds us of the calculus of the length of the hypotenuse of a triangle, according to the theorem of Pythagoras: the square of the distance in space-time is the sum of the square of the space distance and of the square of an imaginary spacelike distance, that of the distance traveled by the light in the considered time interval multiplied by the number i. Therefore it is not the time that is spacelike in Einstein's relativity theory but the time multiplied by i—which corresponds to an imaginary time.

Despite the assaults of physics on the weirdness of the fourth dimension, time still remains fundamentally different from the other three dimensions (of space). The fourth dimension conserves its aura of mystery, even after Einstein.

CHAPTER SEVEN

THE BOOTSTRAP PRINCIPLE AND THE UNIQUENESS OF OUR WORLD

EDDINGTON AND THE EPISTEMOLOGICAL PRINCIPLES

The English astronomer and physicist Arthur Stanley Eddington (1882–1944) is the first scholar who was able to reformulate the old problem regarding the possible uniqueness of our world in modern terms.

The origin of Eddington's s reflection lies in the recognition of certain laws of modern physics, such as Heisenberg's uncertainty relations, as laws with an absolutely novel character, authentic *epistemological principles*. This observation, in itself, is not absolutely original. The uniqueness of Eddington's thinking is manifested in the relationship that he postulates between these epistemological principles and the observer.

Eddington was convinced that all the fundamental assumptions that determine the development of physical theories could be replaced by epistemological principles. "All the usual laws of nature, considered as fundamental," wrote Eddington, "can be entirely provided by epistemological considerations."[1] Epistemological principles are therefore more general than the fundamental laws of nature: *a very limited number of epistemological principles can explain a large number of fundamental laws of nature*.

According to Eddington, epistemological principles themselves can be deducted based on *the sensory and intellectual knowledge of the subject*. To the extent that this structure is unique, reality itself should be unique. The observer is integrated into the reality that he or she observes.

Eddington's ideas are summarized in what is called *Eddington's principle*[2]: all the fundamental statements of physics can be deduced using logical reasoning based on qualitative statements: thus, pure numbers of physics that are *constant* (such as the constant of the fine structure, which characterizes electromagnetic interactions) can be derived without recourse to any experimental data.

Of course, this principle is concerned with the most general aspects of nature, those related to the *invariance* of physical laws: the radius of the atom or the radius of Earth would never be deducted, taking into account epistemological principles.

The assertion that Eddington's ideas demonstrate an extreme idealism is false. The outside world remains more necessary than ever: "We refuse to admit," wrote Eddington, "the probability that the outside world, despite the difficulties we face to get to it, could be disqualified, because it would not exist. . . . The outside world is the one examined by our common experience, and for us, no other could play the same role."[3]

It is well-known that Eddington's ideas were very poorly received, both by philosophers and by physicists, who saw Eddington's approach as a consequence of an intolerable confusion between philosophy, epistemology, and science: how could one accept the idea that epistemology becomes the engine of science if one takes into consideration the most general and fundamental parts of it?

The implicit question, which is the core of Eddington's reflection, still remains of great current interest: could the arbitrary be tolerated in a theoretical description of reality—whether it is about concepts (Why are there three dimensions of space and one dimension of time? Why is there a well-defined number of quarks?) or about pure constants (Why has the constant of fine structures the value it has?)? The fact that scientific experiments show the validity of a particular scientific concept or the numerical value of a certain constant is not an answer to our question: we want to know the *what for* about things. In fact, the progress of a scientific theory is related to the elimination of arbitrariness from the scientific description: a theory based on a small number of axioms is rightly considered far more interesting than a model tolerating hundreds of concepts and arbitrary parameters in order to describe the same set of phenomena. Some physicists believe that a certain degree of arbitrariness will always exist, but there is also the logical possibility of a total elimination of the arbitrary, and as we have seen, this is present in the M theory.

UNITY AND SELF-CONSISTENCY: THE BOOTSTRAP PRINCIPLE

The bootstrap hypothesis emerged first as a possible explanation of certain experimental data in particle physics. This hypothesis was formulated for the first time by Geoffrey Chew, a professor at the University of California, Berkeley,[4] in 1959 and was immediately used for detailed physics calculations by Chew and Mandelstam.[5] The word *bootstrap* itself is untranslatable.

Indeed, *to bootstrap* also means "to pull oneself up by one's bootstraps." The most appropriate term in translation would be *self-consistency*.

The bootstrap theory has emerged as a natural reaction against classical realism, which received its death blow, and against the idea of a need for equations of motion in space-time, an idea with which it was associated during the formulation of quantum mechanics around 1930.

We learned about the existence of Newtonian equations of motion in order to describe physical reality: Newton's equation regarding macroscopic bodies, Maxwell's equations for electric and magnetic fields, and Schrödinger and Dirac's equations for the movements of atomic systems. The movement described by these equations is that of certain entities considered as fundamental building blocks of physical reality, defined at each point of the space-time continuum. By definition, these equations possess an intrinsic deterministic character (the fact that in some cases large ensembles of objects can lead to chaotic behavior does not alter the deterministic character of the basic equations of motion).

Quantum entities are not subject to classical determinism. The bootstrap theory is just drawing the logical conclusions of this situation by proposing the abdication of any equation of motion. This attitude is consistent with the schedule of the matrix S ("S" is the initial for the English word *scattering*) initiated by Heisenberg in 1943[6]: a realist theory must be expressed in terms of quantities directly related to experimental observation.

The abdication of any equation of motion has an immediate consequence: the absence of any fundamental brick of physical reality. "After all," writes Chew, "if one accepts the need for an equation of motion, this means that there is a certain fundamental entity that is in motion; this cannot be a particle—it must be something different, maybe a field. As I noted, there might be *no* fundamental entity, field, or something else."[7]

Here is a more precise definition of the bootstrap, given by Chew: "The only mechanism that meets the general principles of physics is the mechanism of nature; . . . the observed particles are the only quantum and relativistic systems that can be designed without internal conflict. . . . Each nuclear particle plays three different roles: 1) *constituent* of the compounds; 2) mediator of the force responsible for the overall cohesion of the compound; 3) *composed* system."[8]

In this definition, the part appears, simultaneously, as the whole. Nature is conceived of as a global entity, *inseparable* at a fundamental level. The particle plays the triple role (as listed above) of a *system* in irreducible interaction with other systems, which is a first rapprochement between the bootstrap theory and the current systemic thinking.

What is in question in the bootstrap theory is actually the notion of a precise identity for a particle, substituting the notion of a relationship between events, the *event* signifying the context of the creation or annihilation of some particles. Responsible for the emergence of what is called a particle are the *relationships* between events. There is no object in itself, possessing its own identity, which can be defined as separate or different from other particles. A *particle is what it is because all other particles exist simultaneously*: the attributes of an entity that is physically determined are the result of the interaction with other particles.

The bootstrap is therefore a vision of the world's unity, a principle of nature's self-consistency: *the world built on its own laws through self-consistency*. It represents a break with the scientific thinking based on the tradition of Democritus and Newton: the concept of fundamental (and thus *arbitrary*) entities characterizing material *substance* is replaced by a concept of the organization of matter—that of self-consistency. It should be noted that the bootstrap principle is both *organizational* and *structuring*: an *infinite* number of self-consistency conditions determine the existing particles in a *unique* way.

The only world that is compatible with the laws of nature (which can themselves be deduced, in principle, through self-consistency) is, according to the bootstrap principle, the world of nature: it is impossible to find a system without internal contradiction on the logical plane and that is at the same time in agreement with everything that is observed or will be observed. In other words, we can only make a series of approximations. It is a confirmation of modesty when facing the rational order of nature but also a statement of hope.

The bootstrap has important implications for the nature of scientific prediction. The knowledge of the whole claims a long and patient investigation. Therefore, it must be admitted that one looks for ways of approaching self-consistency with the conviction that behind the approximation hides a fundamental coherence, a rational order without gaps: once we obtain partial information of the real world, knowledge of the rest of the world is not arbitrary—it is obtained by self-consistency.

Obviously, there are different degrees of generality in the formulation of the bootstrap's principle.

Thus, the conception of a postulation of a very general form of the bootstrap principle is very logical, including not only particles but also macroscopic bodies, life, and even consciousness: the self-consistency of the whole requires the inclusion of all aspects of nature. Under this very general form, the bootstrap principle has, with the current state of knowledge, an *unscientific* character. This fact was emphasized by Chew himself in an article published in 1968: " 'Bootstrap': A Scientific Idea?"[9] Thus, we have to make a difference between the entire, unscientific bootstrap and the partial, truncated bootstrap, which can be scientifically productive and effective. Such a partial bootstrap—the *hadron bootstrap*—is the one that enjoyed the interest of physicists in the decade 1960–1970.

IS THERE A NUCLEAR DEMOCRACY?

The partial bootstrap proposed by Chew in 1959 refers to the world of hadrons (for example, the proton or the neutron) that have a strong interaction exercised at the atomic nucleus scale. Most of the currently known particles are included in this category. Other particles, such as leptons or photons, are excluded from the hadron bootstrap, which is already an inherent limitation of the approach.

The hadron bootstrap is formulated within the general framework of the S matrix theory.[10] The essential idea of this theory is that of focusing efforts on the scientific description of hadrons as they appear *before* and *after* their interaction. An element of the S matrix describes a specific reaction between particles that depends on the quantity of motion and on the initial and final hadrons' spins (the *spin* is the intrinsic angular momentum of the particle). What happens at the very moment of the interaction is not described. Thus, it is not necessary to introduce a space-time at a microscopic scale: macroscopic space-time is sufficient for the experimental definition of the quantities of motion and of the hadrons' spins. In particular, it is not necessary to define the position of a hadron.

These elements of the S matrix describe all the reactions that can be designed between hadrons. They represent the amplitudes of probability regarding the probability that a given set of measures in the final state comes next after a given set of baseline measures. These amplitudes of the probability of transition directly correspond to measurable quantities.

A true theory must therefore be able to predict the ensemble of these elements of a matrix that is the S matrix itself. The hadron partial bootstrap postulates that it is possible that a single S matrix could be predicted based on some very general restrictions. The axiomatic meaning of these constraints determines the partial character of the hadron bootstrap; in turn, it should be possible, in principle, to deduce those restrictions by self-consistency.

The hadron bootstrap takes into account four general restrictions, all based on extensive experimental evidence.

In the hadron bootstrap theory, all the restrictions imposed play the role performed by the equations of motion inside the theories that are part of the Newtonian tradition. In fact, a set of coupled, nonlinear equations, infinite in number, is generated. It is possible for such a system of equations, which is highly restrictive, to generate a *unique solution* of the hadrons' bootstrap. But the problem is extremely complex on the mathematical plane, and it has not been possible to solve this system of equations; therefore, it is not even known whether the system has a solution (which does not necessarily have to be unique). Subtle and refined mathematical methods, adapted to completely solve the bootstrap problem, have simply not yet been invented.

We have had to satisfy ourselves with the formulation of models that try to take into account the restrictions of the hadron bootstrap as much as possible. These models have been important for both theory and experimental designs. Thus, one of them, Veneziano's model, has generated a whole line of research—that of superstrings (see chapter 5). The agreement of certain models of the hadron bootstrap with experimental data was impressive, yet physicists have moved away from the theory of the hadron bootstrap.

The mathematical complications inherent in the bootstrap program do not represent the only explanation for this decline. Apparently, an irreconcilable opposition separated the hadron bootstrap, together with its nuclear democracy, from the theory of quarks. On the other hand, encouraged by the experimental success of unified theories, scholars moved their center of interest toward the theories of the unification of all physical interactions. By definition, the hadron bootstrap included only the strong interactions.

The bootstrap had to leave the physics scene as a means of calculating physical interactions; however, the bootstrap principle still remains alive. It is referred to in the most diverse contexts, whenever the self-consistency of a theory is required: "With regard to dynamics, when everything else fails, our Supreme Court of Appeal is the bootstrap mechanism, the principle of the self-consistency of the Universe," writes Abdus Salam.[11]

THE BOOTSTRAP AND THE ANTHROPIC PRINCIPLE

Apparently, there is no relationship between the bootstrap theory, formulated in particle physics, and the *anthropic principle*, postulated in cosmology. A subtle and significant connection can nevertheless be established between the two approaches.

Let us recall some important aspects of the anthropic principle.

The anthropic principle (*anthropic* comes from the Greek word *anthropos*, meaning "man") was introduced by Robert H. Dicke in 1961,[12] two years after the formulation of the hadron bootstrap. Its usefulness was demonstrated by the works of many scientists.

Today there are different formulations of the anthropic principle. Despite their diversity, one may recognize a common idea that crosses all these formulations: the existence of a *correlation* between the occurrence of humanity, of intelligent life in outer space (and therefore on Earth, the only place where we can identify this intelligent life), and the physical conditions governing the evolution of our universe. This correlation seems to be subject to strong restrictions: if you change the value of certain physical constants or that of the parameters occurring in some places even a little, then the physical, chemical, and biological conditions that allowed the appearance of humanity on Earth are no longer met.

The existence of these strong restrictions could lead to the hasty conclusion (already drawn by certain scientists and philosophers) that our universe was designed in order to allow the emergence of intelligent observers—us—according to a predetermined project. This conclusion is too simplistic and crumbles in front of a finer analysis concerning the anthropic principle in cosmology.

First of all, Dicke takes some numerical coincidences observed around 1930 as a starting point for the formulation of the anthropic principle. In particular, Dirac noticed the existence of certain empirical relationships between some important constants when they are expressed in a dimensionless form (that is, in the form of pure numbers that are not linked to units of measure). Thus, for example, the age of the universe expressed in atomic units, the number of non-zero mass particles (such as protons) in the visible region of the universe, and a certain dimensionless shape of the gravitational coupling constant may be taken into account. Dirac noticed that all these constants are expressed as integer powers (positive or negative) of the same number (10 to the power 40, or 10^{40}), suggesting that these coincidences may be the manifestation of an unknown law of that age. It is also interesting to note that, for reasons that have to do with the logical consistency of his suggestion, Dirac was forced to postulate that certain constants, such as that of gravitational coupling, must *vary over time*. Referring to Dirac's suggestion, Dicke made the fundamental observation that *the age of the universe cannot have an arbitrary numeric value* when human existence is taken into account in its physical, chemical, and biological constitution: it must be that those non-uniformities, which are the galaxies and which give rise to stars and planets, take the time (and the necessary conditions) to compose themselves; it is also required that the heavy elements that make up our bodies and are needed for life have time to compose themselves through stellar nucleosynthesis.[13]

Some strong constraints seem to be exerted not only on the age of the universe but also on other physical and astrophysical quantities. Brandon Carter underlined the role of the gravitational coupling constant, which had to be close to the experimentally observed value so the planets could appear and survive. A gravity that is too strong or too weak leads to the impossibility of the formation of planets. The coupling constant characterizing strong interactions is also very constrained. According to Frank Tipler,

> If the strong force had been slightly stronger than it is, then it would have been possible for helium to form with only two protons. . . . These protons would be able to maintain themselves

together in a normal way, but the electromagnetic force would be unable to reject two foreign protons, and it would therefore become very easy, for this reason, to trigger a nuclear response by only lighting a match; ... in the early universe, all the hydrogen would have been converted to helium, and there would have been no hydrogen available for the formation of the stars of greatest magnitude.... If, on the contrary, the strong force had been weaker, then complex atoms, such as carbon, could not exist. The rejecting electromagnetic force would tear the nucleus.[14]

In fact, a vast *self-consistency*, which concerns both physical interactions and the phenomena of life, seems to govern the evolution of the universe. Galaxies, stars, planets, humanity, and the atomic and quantum worlds seem united through one and the same self-consistency. The relationships emphasized by the anthropic principle are, in my opinion, a sign of this self-consistency. Again, they raise the problem of the uniqueness of the observed world.

It is precisely through this self-consistency that a link between the general bootstrap principle and the anthropic principle might be determined. It might even be argued that the anthropic principle is a special case of the bootstrap principle.

Certain formulations of the anthropic principle have a strong flavor of finality, which is unacceptable to the modern scientific spirit. As Stephen Hawking noted, despite his contribution to the deepening of the anthropic principle:

The anthropic principle is not completely satisfactory, and we cannot stop thinking that there should be a deeper explanation. In addition, this principle cannot account for all the regions of the universe.... Perhaps our entire galaxy has been molded this way, but that doesn't mean other existing galaxies were necessarily formed the same way, not to mention a few billions of them that we have been spotting almost uniformly throughout the observable universe. The large-scale homogeneity of the universe makes it very difficult to adopt an anthropocentric point of view or to believe that the structure of the universe is determined by something as peripheral as a complicated molecular structure, present on a single planet, spinning in the orbit of an average planet, in the outer suburbs of a fairly typical spiral galaxy.[15]

Even if human beings are, in a sense, a peripheral phenomenon, they seem to be necessary for the self-consistency of the whole. Human beings are not the center but rather a link in the reality that includes them, participants in the dynamic, evolving structure of the universe: the center is everywhere. This is the interpretation that we could offer if the anthropic principle is placed inside the perspective of the general bootstrap principle. The finality (which passes through human beings but is not limited to them) is not preexisting: it is built by the universe itself.

It is also instructive to put in parallel the bootstrap theory in physics and the anthropic principle, as it is applied to cosmology, from the same perspective of a general self-consistency. The anthropic principle can only be strengthened by an increasingly more detailed dynamic modeling. It is not unlikely that the next decades could witness the birth of a new formulation of the bootstrap principle that would take into account the phenomena of life. Thus, it would be gradually understood whether life is a necessary, nonaccidental phenomenon required by the self-consistency of our universe.

METHODOLOGICAL CONSIDERATIONS

Ever since its formulation, the bootstrap principle has had bitter detractors and passionate advocates.

Thus, in 1965, three Japanese physicists belonging to the materialist-dialectical school in Nagoya (formed around Sakata and Taketani) wrote, "such trials will lead us to Leibniz's philosophy, which conceives the universe as having a predetermined harmony. This view will introduce religious elements in science and will stop scientific thinking at this level."[16] Although giving full credence to neither the ideas of Nagoya's school nor the idea of the bootstrap, the Israeli physicist Yuval Ne'eman wrote in 1975: "we see now, at Berkeley and elsewhere, another attempt at describing the hadron matter, followed by an almost equal dogmatic narrowness, its fundamental motivation being the belief that we have reached the end of the road. This is the bootstrap current."[17] Echoing these words, in 1977 the journalist Nigel Calder wrote: "The bootstrap . . . involves a conscious rejection of the objective of traditional physics which is to explain events in terms of forces that act between well defined particles. . . . The particles themselves cease to be an object of analysis; they are rather relationships between events. If Chew was right, this would mean very bad news for Western philosophy and science, with their aim to banish the redundant mystery of the universe, revealing its fundamental entities and laws."[18] A similar, very plausible controversy could be produced with regard to superstrings.

The first novelty of the bootstrap principle is, as physicist James Cushing rightly notes, of a methodological nature.[19] The hypothetical-deductive method has become increasingly more prevalent in physics over the last three centuries: it presupposes the existence of certain fundamental laws, raised to the rank of axioms, all their consequences being deduced. A scientific explanation should follow this path. Usually, the assumption is that these laws shall be exercised within certain fundamental entities, contingent and independent, but in interaction, the laws are also formulated.

By proclaiming the absence of laws and fundamental entities, the bootstrap principle introduces a radical new methodology into science. It is true that, in practice, it is limited to a partial bootstrap, which tolerates some axioms as a starting point, and so there is a certain resemblance to the standard, hypothetical-deductive methodology. But the refusal to introduce fundamental entities and equations of motion that are associated with them persists in various schemes of the partial bootstrap. Therefore, the bootstrap as it is applied in physics represents a deviant methodology that is only accepted when the standard methodology is in crisis.

This is precisely what happened with the hadron bootstrap, formulated in an era characterized by a tremendous proliferation of various hadrons: in such circumstances, it was impossible to maintain the assumption of fundamental entities among these hadrons. Moreover, if the idea of fundamental entities is old, earlier than the twentieth century, then the idea that particles can be *concurrently* elementary and composed could not have a meaning without the contributions of quantum theory and those of relativity theory. In fact, this is reflected in some important developments in quantum chromodynamics.

The aesthetic attraction of the bootstrap approach, associated with its successes regarding experimental data, made this deviant methodology dominant in the physics of strong interactions in the 1960s. Meanwhile, the standard methodology saw a dramatic reversal when the quarks model was invented and quantum field theory was developed.

Along with physics and its methodology, the bootstrap principle has attracted the attention of philosophers. An immediate temptation is to relate the bootstrap to Leibniz's monadology. Scholarly studies were published in this direction by George Gale.[20] It is still difficult to advance anything other than a few analogies. As a simple substance, without parts, the monad, even if it is a mirror of everything, seems very different from a particle that consists of all other particles. The relationship between monads is static; that between particles is dynamic, based on a continuous exchange of information. In addition, the bootstrap's world is not "the best of all possible worlds"; it is neither the worst nor the best, but simply the only one that can exist by self-consistency. The bootstrap principle seems, in this respect, closer to Anaxagoras, who proclaimed that everything is in everything and that an object is what it is because all other objects exist at the same time (an object simply corresponds to its predominant component): "In reality, nothing is either made, nor destroyed, but is made and separated, based on beings that exist."[21] His doctrine about the *homeomeries* is not foreign to the notion of unobservable subconstituents. Other ideas of Anaxagoras, such as the unity of contradictories (for example, "snow is white, but water is black"), the unity of matter in the universe, and the purpose that is created through the progress of rationality, deserve also to be mentioned in this context.[22]

Another temptation is to approach the bootstrap and Taoism.[23] However, the bootstrap is closer to Anaxagoras, Boehme, and Peirce or to current systemic thinking than it is to Leibniz or Eastern philosophy.

Finally, one last attempt is to see the bootstrap principle as an illustration of Hegel's dialectics.[24] But the way the bootstrap realizes the unity of contradictions is closer to Lupasco's view than it is to Hegel's.

Whatever the fascination exerted by the bootstrap principle, from the philosophical perspective, its interest lies in something other than what impresses physicists. More than a new theme in physics, it is rather a *symbol* opening a window on the world's unity. This symbol is inexhaustible, always remaining the same. Its richness includes a manifestation in the field of natural systems. Certainly there is a *total bootstrap*, which is a view of the world, as well as a *partial bootstrap*, which corresponds to a scientific theory. One without the other is poor and, ultimately, sterile. The view of the world is nourished and enriched with information derived from the natural world and, in turn, scientific theory gets a *human* dimension through the existence of a vision. The major interest of the bootstrap principle is that of unifying a worldview with a scientific theory in one and the same approach.

CHAPTER EIGHT

COMPLEXITY AND REALITY

> True Physics is the one that will succeed, one day, to integrate the whole Man into a coherent representation of the world. . . . Man, not as a static center of the World, as it was long believed, but as an axis and a flagship of Evolution.
>
> —Pierre Teilhard de Chardin, *Le Phénomène humain*

THE EMERGENCE OF COMPLEX PLURALITY

In the course of the twentieth century, complexity—frightful, terrifying, obscene, fascinating, and invasive—has established itself everywhere as a challenge not only to our existence but also to its very meaning. Meaning seems to get absorbed, as if by the white blood cell of complexity, in all areas of knowledge.[1]

This complexity is nourished by the disciplinary research boom, which, in turn, leads to the accelerating proliferation of disciplines.

Classical binary logic confers its patent on either a scientific or nonscientific discipline. Thanks to its rigid norms of truth, a discipline can pretend to contain all knowledge within its own field. If the discipline in question is considered as fundamental, as a touchstone for all other disciplines, its scope is thereby enlarged so that it appears to encompass all human knowledge. According to the classical viewpoint, disciplines as a whole were conceived of as a pyramid, the base of which was physics. Complexity literally pulverized this pyramid, provoking a veritable disciplinary big bang.

The fragmentation of the disciplinary universe is in full swing today. The domain of each discipline is inevitably becoming more and more specific; that which enables communication between disciplines is becoming more and more difficult, even impossible. A multischizoid, complex reality appears to have replaced the simple, one-dimensional reality of classical thought. In turn, a subject becomes a shambles when it is replaced with an ever-increasing number of separate parts studied by different disciplines. This is the price that a subject must pay for a certain kind of self-established knowledge.

There are multiple causes for this disciplinary big bang. The fundamental cause is easy to discern: the disciplinary big bang is the response to the demands of a technoscience without brakes, without values, without any end other than utilitarianism.

But this disciplinary big bang also has enormous positive consequences, because it has led to an unprecedented understanding of the knowledge of the exterior universe, as well as contributing new impetus to the establishment of a new worldview. Because a stick always has two ends, when the weight shifts too far in either direction, the balance must be restored.

In a paradoxical way, complexity is embedded in the very heart of simplicity: fundamental physics. Indeed, popular works state that contemporary physics is a physics where a wonderful simplicity rules through fundamental "building blocks"—quarks, leptons, and messengers of the physical interactions. Each discovery of a new building block predicted by this theory is saluted by a Nobel Prize and is presented as a triumph of the simplicity that rules in the quantum world. But for physicists working inside physics, the situation appears as infinitely more complex.

For example, according to superstring theory in particle physics, physical interactions appear to be very simple, unified, and subordinate to general principles if they are traced within a multidimensional space-time continuum and involve an incredible energy corresponding to Planck's mass. But complexity appears at the moment they pass into our world, which is characterized by four dimensions and by low energies. Unified theories are at their strongest at the level of general principles, but they are very poor at describing the complexity of our own level of reality. There are even some rigorous mathematical results that indicate that the passage from one single unified interaction to the four known physical interactions is extremely difficult, even impossible. A raft of mathematical and experimental questions of extraordinary complexity remains unanswered. In contemporary physics, mathematical and experimental complexity are inseparable.

Complexity is manifested everywhere else in the exact (hard) or human (soft) sciences. In biology and in neuroscience, for example, which are presently rapidly developing, each day brings more complexity and new surprises.

The development of complexity is particularly striking in the arts. By an interesting coincidence, abstract art appeared at the same time as quantum physics. Since then, an increasingly chaotic development seems to dominate more and more in formalized experiments in this field. With some notable exceptions, such as Brancusi's sculpture, meaning has been wiped out by form. The human face, so beautifully represented in the art of the Renaissance, has been increasingly distorted until it seems to have disappeared totally into absurdity and meaninglessness. A new art, computer art, has arisen and gradually has replaced aesthetic works and acts. In art, as elsewhere, a stick always has two ends.

Social complexity emphasizes the complexity that invades all areas of knowledge to the point of paroxysm. The simplistic idea of a just society, founded on a scientific ideology and the creation of the new human being, is unraveling under the pull of a multidimensional complexity. That which remains is founded on the logic of utilitarianism and suggests to us nothing other than the "end of history." Everything happens as if there were no future. And if there is no future, sound logic tells us that there is no present. The conflict between individual life and social life grows with an accelerating rhythm. And how can one dream of a social harmony based on the annihilation of the interior being?

The knowledge of complexity, in order to be recognized as knowledge, bypasses one preliminary question: is the complexity of which we speak a complexity without order, in which case knowledge of it would have no meaning, or does it contain a new order and a new kind of simplicity, which could appropriately become the object of a new knowledge? An authentic attempt to answer this question leads to a possibility somewhere between damnation and salvation.

Is complexity itself created by our minds, or is it found in the very nature of things and beings? The study of natural systems gives us a partial answer to this question: both are true. Complexity in science is, first of all, the complexity of equations and of models. It is therefore the product of the mind, which is inherently complex. But this complexity is the mirror image of the complexity of experimental data, which proliferate endlessly; it is therefore also in the nature of things.

Further, quantum physics and quantum cosmologies show us that the complexity of the universe is not the complexity of a garbage can, without any order. A stunning coherence exists in the relationship between the infinitely small and the infinitely large. A single element is lacking in this equation: our own finite realm, with its infinite dimensions. Humanity remains silenced and estranged in the comprehension of complexity, and with reason, because it has been proclaimed dead. Between the two ends of the stick—simplicity and complexity—the included middle is missing: humanity, the subject.

SOME REFLECTIONS ON SYSTEMIC THINKING

Two conflicting theories, atomistic thinking and systemic thinking, cross the entire history of human thinking. Far from being confined to the study of nature, their antagonism was manifest in the different types of civilizations and societies humanity has seen. This antagonism penetrates even into the intimate life of human beings, who see themselves torn between an obscure desire for cosmic unity, a belief that they are made in the image of a unique God, and a more or less firm belief in their absolute realization that they are individual units.

However, it is possible to recognize a certain asymmetry between atomistic thinking and systemic thinking. Atomistic thinking seems to correspond better to the natural tendency of human intelligence to reduce all reality to its manifestation in the space-time continuum of common experience resulting from our observations at our own scale.

In *De nature rerum*, Lucretius (ca. 98–55 BC) ridiculed the doctrine of Anaxagoras (ca. 500–ca. 428 BC) on behalf of common sense, which claimed that Everything is in everything. In the modern era, the opposition between Anaxagoras's doctrines and those of Lucretius is found in the dispute that marked the 1960s between the bootstrap supporters and the supporters of quantum field theory, as well as in current discussions on the status of superstring theory.

Today's science has managed to highlight the systemic intelligence of the natural world.[2] In this way, new forms of systemic thinking have gradually been able to emerge. They are based on an assumption of *isomorphism* between the different sciences: laws studied by various sciences are very different from each other, but one can guess that there are very broad laws that allow a certain unity in science.

SYSTEMIC THINKING AND QUANTUM PHYSICS

Current systemic thinking has emerged from a rejection of classical realism, which is perceived as incompatible with modern scientific data, and from an attempt to put *complexity* in order.

Indeed, our world seems overrun by complexity. Everywhere we look—to the large infinite, to the small infinite, and even to our own scale—we see complexity manifesting itself and triumphing. "At the beginning there was complexity," writes Edgar Morin.[3] Loyal to analytical thinking, the modern person passes, like a stranger, through a world that is increasingly more unexplainable.

In an attempt to easily understand the emergence of complexity in the natural systems, let us recall an example given by Ludwig von Bertalanffy (1901–1972).[4] Let us consider a set of N points joined by segments with or without an arrow above. This aggregate forms a graph that has itself a certain global orientation (toward the right or toward the left). If there are two points, we have four possible ways to connect them. But it is enough to take, for example, five points for the number of possible connections to exceed one million. We could try to imagine the huge number of connections between the neurons in our brain or between the atoms in the universe. Classical analytical thinking could never exhaust the richness of natural systems, even if everyone dedicated their work to the study of these systems. New methods of thinking are needed in order to address the problems of complexity.[5] Despite the great diversity that characterizes the systemic approaches to reality, some common main ideas may be recognized.

For example, the universe can be conceived as a vast whole, as a vast cosmic matrix in which everything is in perpetual motion and energetically structuring itself. But this unity is not static; it implies differentiation, diversity, the emergence of hierarchical levels, and the occurrence of relatively independent systems, of objects as local configurations of energy.

The different systems are combinations of elements that are in an interaction that can never be reduced to zero: lack of interaction would mean the death, the disappearance of a system, its decomposition into constituents. The very existence of a system means that it is not just the sum of its parts.

The *opening* of the system, through its interaction with other systems, is what prevents its degeneration, its death, by the inevitable degradation of energy, by the growing disorder. This way, *systems of systems* can be put together in order to build up the full diversity of the world in a perpetual and universal exchange of energy, in a vast and ceaseless *nonseparability*, a real rescue for the lives of systems.

Unlike reductionism, which explains diversity through a substance that is common to various systems, systemic thinking speaks of a common *organization*. *Today matter means the complexus energy—substance—space-time—information*. Natural systems are formed by themselves, being created in time. Natural systems avoid balance, which is equivalent to degeneration and death, choosing stability in a state of *imbalance* through openness toward other systems. Thus, fluctuations become the source of *evolution* and *creation*.[6] The self-creativity and self-organization of natural systems are undoubtedly signs of freedom, but this freedom is exercised within the limits of its conformity to and compatibility with the dynamic needs of the whole.

Unity in diversity and diversity through unity—this seems to be the message of natural systems.

It is obvious that these brief considerations cannot illustrate the whole richness of contemporary approaches resulting from fields as different as biology, economics, chemistry, ecology, and physics. We have referred here only to the emergence of systemic issues in modern physics.

The new concept of matter in today's physics is a sign of systemic coherence. The postulate of universal interdependence underlying systemic thinking (at least for open systems) is very close to the bootstrap principle. The self-organization of natural systems is a concept that significantly overlaps that of natural self-consistency. The concept of spontaneously broken symmetry manages to reconcile the unity and the diversity of physical interactions. Finally, the role of quantum fluctuations in the structure of matter is obvious, if we consider the concepts of quantum vacuum and superstrings. One could give many other similar examples.

However, paradoxically, quantum physics plays a very small part in current systemic thinking, with the exception of some considerations of its random quantum indeterminism. For example, the voluminous work of Erich Jantsch (1929–1980) on the study of the self-organization of the universe[7] devotes only a few pages to particle physics and its relationship with modern cosmology.

The knowledge of the laws operating on a quantum scale leads to the development of a true systemic methodology.

LEVELS OF REALITY

The notion of complexity seems to be inherently ambiguous in systemic thinking.

The first source of this ambiguity is the fact that there are basically *two types of complexity*: one that describes the transition from one level to another and one describing complex phenomena on a determined scale. The confusion between the two types of complexity leads to endless misunderstandings. If it is true that organized complexity implies an increased scale, then *gigantism* should correspond to maximum organization and interaction, which is obviously false. On each scale there is an optimal *size* of systems, starting from which organization and interaction begin to decline. The damages created by gigantism are perceptible in the economic and social fields.

A second source of ambiguity comes from the dependence of *complexity on the nature of space-time*. In most systemic studies, complexity is implicitly associated with the continuous space-time with four dimensions that characterizes our own scale. But continuous space-time with four dimensions is not the only one designed to describe natural systems. As we have seen in chapter 5, a space-time with more than four dimensions or even a discontinuous space-time could be conceived on a quantum scale. The complexity will be of a different nature in each case. As long as the macroscopic world can be derived starting from the quantum world, the understanding of the complexity that operates on our scale can be examined further by clarifying its relationship with other types of complexity.

These uncertainties, these ambiguities disappear when the concept of *levels of Reality* is introduced.

Simultaneously, we give its pragmatic and its ontological meaning to the word *reality*. By *Reality* (with a capital R), we intend, first of all, to designate the reality that *resists* our experiences, representations, descriptions, images, and mathematical formulations.

We also need to give an ontological dimension to the concept of Reality, to the extent that nature participates in the world's existence. Reality is not only a social construction, a community consensus, an intersubjective agreement. It also has a *transsubjective* dimension, to the extent that a simple experimental fact can frustrate the most beautiful scientific theory.

Of course, we must distinguish between *Real* and *Reality*. *Real* means *what it is*; Reality is related to the resistance of our human experience. The Real is, by definition, always hidden. Reality is available to our knowledge.

By *levels of Reality*,[8] we must understand a set of systems that is always *invariant* under the action of a number of general laws (in the case of natural systems) or of a number of general rules and norms (in the case of social systems): for example, quantum entities are subject to quantum laws, which are radically different from those of the macrophysical world; as another example, individuals are subject to general rules and norms that are radically different from those for society. Two levels of Reality are *different* if, while an individual passes from one level to another, there is a rupture of laws and a rupture of fundamental concepts, such as causality (in the case of natural systems), or there is a rupture of general rules and norms, such those governing the spiritual life (in the case of social systems).

The occurrence of at least three different levels of Reality in the study of natural systems—the macrophysical level, the microphysical level, and cyber-space-time (on top of which we can add a fourth one, at the moment purely theoretical, of superstrings, considered by physicists to be the ultimate skeleton of the universe)—is a crucial event in the history of knowledge. This fact can make us rethink our individual and social lives, undertake a new reading of the old knowledge, and explore ways of knowing ourselves here and now.

A flow of information is transmitted in a consistent manner from one level of Reality to another level of Reality of our physical universe. As an example of this kind of vertical exchange, it can be assumed that the deep significance of Heisenberg's uncertainty relations is connected with the possible exchange of information between the world of particles and other systems of a different scale. The probabilistic nature of quantum events could thus acquire an interesting explanation.

Consistency is *oriented*: there is an arrow associated with the transmission of information from one level to another. Therefore, if it is limited only to the levels of Reality, coherence stops at the "highest" level and at the "lowest" level. If we want an open unity of coherence to exist beyond these two limit-levels, we must consider that the set of Reality levels is extended by an area of nonresistance to our experiences, representations, descriptions, images, and mathematical formalities.

The nonresistance area of absolute transparency is simply due to the limits of our body and our sense organs, regardless of the ability of instruments to extend these sense organs. The nonresistance area corresponds to the *area of the sacred, to that which is not subject to any rationalization*. It is worth remembering the important distinction made by Edgar Morin between rational and rationalization[9]: the sacred is rational, but it is not rationalizable.

The problem of the *sacred*, understood as the presence of something in the area of the *irreducibly real* in the world, is inevitable in any rational approach to knowledge. The presence of the sacred in the world and in ourselves can be denied or affirmed, but we are always obliged to refer to the sacred in order to develop a coherent discourse on Reality.

The open structure of all levels of Reality conforms to Gödel's theorem on arithmetic. Gödel's theorem has a considerable significance for any modern theory of knowledge, because it not only refers to the field of arithmetic but also to any mathematics including arithmetic. Gödel's structure of all levels of Reality, which is associated with the included third (see the following chapter), involves the impossibility of building up a complete theory in order to describe the transition from one level to another and, a fortiori, to describe all levels of Reality. If there is a unit linking the levels of Reality, it must necessarily be *an open unit*.

The ensemble of levels of Reality and its complementary zone of nonresistance represent the *object of transdisciplinarity*.

A new *principle of relativity*[10] is generated by the coexistence of complex plurality and open unity: no level of Reality is a primary location from which to understand all other levels of Reality. A level of Reality is what it is because all other levels exist at the same time. This principle of relativity is the beginning of a new vision of culture, religion, politics, art, education, social life. And when our vision of the world changes, the world itself changes.

In his philosophical writings, Werner Heisenberg approached the concept of levels of Reality. In his famous *Manuscript of 1942* (published in German only in 1984 and in French in 1998), Heisenberg, who knew Husserl well, introduced the idea of *three regions of reality* capable of allowing us access to the very concept of reality: the first region is that of classical physics, the second that of quantum physics and of biological and psychological phenomena, and the third that of religious, philosophical, and artistic experiences.[11] This classification has a subtle foundation: that of the growing closeness between subject and object.

Levels of Reality are radically different from levels of organization as they have been defined in systemic theories. Levels of organization do not imply a breach from fundamental concepts: several levels of organization can belong to one and the same level of Reality. Levels of organization correspond to different ways of structuring the same fundamental laws. For complexity is changing its nature. It is no longer a complexity that is directly reducible to simplicity. Different degrees of materiality correspond to different *levels of complexity*: the extreme complexity of a level of Reality can be seen as simplicity when compared with another level of Reality, but the exploration of this second level reveals that this one is of an extreme complexity in relation to its own laws. This structure in degrees of complexity is intimately related to *the Gödelian structure of nature and knowledge* induced by the existence of different levels of Reality.

It is clear that the concept of levels of Reality will have consequences not only on the epistemological plane but also in scientific research, by formulating new problems and areas of exploration. Of the utmost importance is the fact that this concept has proved to be crucial in the contemporary dialogue between science and religion and, more generally, between science, culture, spirituality, and society.

IS THERE A COSMIC BOOTSTRAP?

How does the universe function? Is it just a sort of machine—of course a spectacular one, but after all, a machine, composed of practically independent systems, mechanically linked to each other? Or rather, does it have an underlying unity, provided by a dynamic intelligence, continuously evolving, functional at every level of nature? Are there laws that run across all scales of nature (particles, atoms, people, planets, and so on), invariable laws that still have different effects, depending on the scale on which they function? In other words, is there an interconnection between the different scales of nature, or is the universe just a sad machine, each scale being destined for destruction and death by the continuous growth of disorder, of entropy?

Obviously, these are difficult questions that humanity has always asked in one form or another, questions for which we cannot have definitive answers now, because our view of nature is in continuous flux. However, what is new in our age is the considerable experimental and theoretical progress of modern science, which has managed to penetrate the very core of the matter rigorously and mathematically. If the above questions are metaphysical by their very nature, they can still find a partial, although penetrating, clarification through the rational approach of science. These questions are no longer confined by the human subjective experience but essentially acquire new characteristics by finding correspondences in terms of matter, of the material universe in which we are immersed.

An example of such a development of modern science is the birth of a new, truly interdisciplinary branch of science—*quantum cosmology*. As its name indicates, this new science is based on *the idea of unity* between the two scales of nature that were considered to be totally different only a few years ago—the quantum scale and the cosmological scale. The interactions between particles can enlighten us on the evolution of cosmos, and the information about cosmological dynamics can clarify certain aspects of particle physics.

The idea of unity between the quantum scale and the cosmological scale is, after all, as strange as it seems at first. According to the big bang theory, the universe must have been a tiny size in the beginning. At that time (10^{-42} seconds after release), quantum processes must have been important and even prevailing. It is normal for the universe to retain a memory of what happened at its birth.

In quantum cosmology, issues considered taboo for science in the recent past can be addressed by means of scientific inquiry. Can we imagine a cause for the big bang? Can we talk scientifically about an origin of the universe? Where does the extraordinary energy consumed during the big bang come from? Even an issue considered to be metaphysical in nature (for example, the genesis of space and time) can be challenged by science today "at least logically," says Steven Weinberg. "It might have been a start, and time itself might have had no significance before that; we may have to get used to the idea of an absolute zero of time—a moment in the past beyond which, in principle, it is impossible to imagine a chain of causes and effects."[12]

Of course, the reader should not imagine that today's science has definitive answers to such questions. Quantum cosmology is an evolving field, particularly due to superstring physics. Different approaches are continually modified under the pressure of the necessity for self-consistency in both the theoretical and experimental plane. But the *entry* of such issues into the field of scientific research seems to be a definitive achievement.

I shall give only one example.

If we do not take into account the gravitational force (of attraction), it is impossible to understand the nature of the initial explosion (the big bang), an event associated with the manifestation of a considerable force of rejection that led to the expansion of the universe: the big bang must be accepted as a postulate, as an unexplainable initial condition. But if one takes into account the unification of all physical interactions, the new unified interaction can generate the force of rejection that characterized the big bang. In 1981, a scenario that could explain the big bang (a scenario called *inflation*) was proposed by the American physicist Alan Guth.[13] The original scenario of inflation that he proposed suffered some changes, but the basic ideas remain the same.

The essential idea of this scenario is related to the recognition of the fact that the quantum vacuum does not have only one state but several possible states: a state of true vacuum ("true" in the sense that it corresponds to the lowest possible energy) and states of false vacuum ("false" in the sense that it has a higher energy than true vacuum). The states of false vacuum can be separated from the state of true vacuum, as in the quantum theories of grand unified fields, through considerable energies. If the universe is generated through a false vacuum state, it will have a very rapid period of exponential expansion; in other words, it will experience a veritable inflation. At the end of the inflationary period, just as with any excited state, the false vacuum state decays (moving toward the true vacuum), thus releasing a significant amount of energy. It is precisely this energy that would be at the origin of the big bang.

Even if one accepts that inflation was somehow the cause of the big bang, we can wonder where the considerable energy of the inflation itself came from. A possible answer can be obtained by calling on the complex properties of the quantum vacuum. The quantum vacuum behaves like a huge battery. A false vacuum state can literally pump energy to the quantum vacuum.

If you are curious by nature, you could push the question even further: where does our space-time come from? An interesting possibility is suggested by superstring theory. If the universe had been associated from its beginning with a multidimensional space-time—for example, a dimension of time and ten dimensions of space—then the spontaneous compaction of seven dimensions of space (namely, their rapid wrapping in an infinitesimal region of space) might have been linked to inflation in the usual three-dimensional space. The seven extra dimensions will remain forever hidden, unseen, but their relics would be precisely the known physical interactions. Why do seven hidden dimensions remain and not six, five, or none? In superstring theory, an interesting argument is given in this regard. Superstrings are wrapped around the spatial dimensions that, early in the beginning of the universe, have the size Planck calculated, and all dimensions may remain hidden forever. These wrapped superstrings prevent these dimensions from increasing. Therefore, unwrapped superstrings must be produced, which could happen with the encounter of a superstring and its antisuperstring. But the probability of this meeting is very low, beyond the three dimensions.

Obviously, quantum cosmology suggests the idea of the *spontaneous emergence* of the universe as a result of physical laws. The universe seems able *to create itself* and also to self-organize, without any outside intervention. The most appropriate image for visualizing this autoconsistent dynamic of the universe would be that of the *Ouroboros*—the snake that bites his tail—an ancient gnostic symbol and also a symbol of the achievement of the Great Alchemical Work.

The ensemble of self-creation and self-organization processes of the universe has been named the *cosmic bootstrap* by the English physicist Paul Davies: "the universe fills itself, exclusively from within, thanks to its physical nature, with all the energy needed to create and animate matter, thus supplying its own explosive origin. This is the cosmic *bootstrap*. We owe our existence to its amazing power."[14]

But how far can we go with the self-organization of the universe? Does humanity have a place in the cosmic order?

EVOLUTION AND INVOLUTION

Besides experiencing a great feeling of wonder when contemplating the unmistakable coherence of the physical universe, we cannot prevent ourselves from simultaneously experiencing a deep feeling of unease. What is the use of all this consistency? What is the use of all these extremely fine and accurate matches between the various physical parameters so that the universe is as it is? For example, it would have been sufficient if the percentage of expansion of the universe was very slightly smaller or bigger than the one observed, for the universe to collapse and completely disappear. At the same time, a high level of consistency is required so that the uniformity of the distribution of matter, on a large scale, experimentally observed, can be held in time. The cooperation between the quantum scale and the cosmological scale is able to provide such fine matches. But all this just to achieve the certain death of the physical universe either by gradual cooling (in the case of an open, ever-expanding universe) or by gradual warming (for a closed universe, which will eventually end by shrinking continuously)?

Nature is supremely ambiguous. Recent experimental findings seem to favor the scenario of a flat universe in accelerated expansion. In order to achieve this expansion, neither the visible substance (of which we are made) nor the substance called "black" (which does not emit light and does not interact with the visible substance) is sufficient. We also have to postulate the existence of a *dark energy* whose nature is not yet known by anyone. So many mysteries in order to reach a foreign universe composed of a different matter from ours, where we will be ever lonelier, ever more distant from other cosmic concentrations.

This feeling of anxiety is expressed well by Steven Weinberg, when he writes "It is almost impossible for human beings to believe that there is no special relationship between themselves and the universe, that life is not just the grotesque result of a series of accidents that go back in the past to the very first three minutes and that we have been specifically designed from the beginning. . . . It is even harder to understand that this universe has evolved from such unfamiliar starting conditions that we can hardly imagine them and that it should end up by extinction, either in endless cold or infernal heat. The more the universe seems comprehensible, the more absurd it also seems."[15]

Is the universe absurd? That may be the case if we overlook the role of life and the human being. We must recognize clearly that we currently have no scientifically based idea of the possible cosmological role of life, although it is logical to imagine that life may play a cosmological role. These days, to express well-defined opinions on the cosmological role of life would mean science fiction rather than science. For now, one could approach a less ambitious issue: that of the place of the human being as a natural system.

A remarkable consistency appears to govern the quantum scale and the cosmological scale. Does the intermediate scale, the scale of the human being, miss the ever more noticeable unity of the world? Violence, lawlessness, lack of consistency: is this precisely the human fate, in contrast with the self-organization and the self-consistency that seem to prevail in other natural systems?

Let us take the existence of a *unity* of natural systems, whatever their scale, as a working hypothesis. Then we could take the dynamic of other natural systems as a *model*, certainly not through naive and brutal transposition, but through an understanding of what ensures the unity of the physical world at the finest level.

An important difference between human beings and most other natural systems is their ability to choose. A car does not choose between moving and not moving. An apple does not choose whether or not to be eaten. Even the universe does not choose among many possible developments. The only exception, in the current state of our knowledge, happens at the quantum scale, which is characterized by a fundamental *spontaneity*, the location of a certain freedom. An event is *unexpected*, nonpredetermined, only on a quantum scale; in this sense, we can speak about a certain freedom.

But between quantum freedom and human freedom there is an important distinction. Quantum events do not choose whether to cooperate or not with one another—their nonseparability forces them to comply with such cooperation. Overall, the quantum scale does not choose whether to cooperate with the cosmological scale. Everything happens as if self-organization and self-consistency could operate *up to a certain extent*, that of the appearance of life and the human being. From now on, *a new kind of choice appears:*

the one between evolution and involution. Human beings seem to be the only natural system that has the possibility to choose between the evolution of their species and its total destruction.

In this way, we can better understand systemic visions of the world, as, for instance, the systemic vision developed by Teilhard de Chardin.[16] Teilhard de Chardin believed, as Frank Tipler does,[17] that humanity is necessarily going toward the *Omega Point:* we have a choice between our progress and our destruction. In the tangled complexity of today's world, it is hard to imagine how an orientation that would guide this choice might appear.

If new phenomena such as the rapid development of the Internet and the awareness of economic interdependence can be considered encouraging signs of a gradual manifestation of *the nonseparability of the humanity body*, they are still too tainted by their commercial aspects to be understood in their whole significance.

The information obtained from the study of natural systems and their integration into a coherent vision of the world could lead to an urgent and necessary transformation of our attitude toward reality. Thus we may start a new era of knowledge, during which the study of the universe and the study of humanity would support each other.

CHAPTER NINE

THE HUMAN BEING

The Most Perfect of All Signs

> The third is what throws a bridge over the gap between the first and last absolutes and links them.
>
> —Charles S. Peirce, *Écrits sur le signe*

NATURAL LANGUAGE AND SCIENTIFIC LANGUAGE

Mathematical thinking has an axiomatic and deductive structure. Axioms are stated, definitions are formulated using specific symbols that are defined once and for all, and then demonstrations are made without introducing any foreign or ambiguous element. The properties of scientific, mathematical language have been studied widely.[1] We are going to recall some of them here.

Scientific language possesses an infinite synonymy—for each sentence, there is an infinite number of equivalent sentences (which have the same meaning), although the synonymy is usually missing from natural language. Conversely, homonymy is usually missing from scientific language: meaning is independent of the one who perceives the mathematical expression. Scientific meanings therefore have a *discrete* nature, in contrast to the continuous nature of meanings that belong to natural language.

Scientific significance is invariant in space and time, which ensures its property of being translatable. Natural language is untranslatable in the strict sense because of its cultural content. Scientific meaning may be communicated from one person to another; it is independent of context. It can be designated true or false; natural language is beyond this opposition.

Finally, scientific meaning is independent of its expression, of its musical structure. Mathematical words thus establish a true dictionary that excludes circular definitions. Conversely, the definitions found in a dictionary of natural language are often circular in nature. The distinction between axiomatic words and derived words does not exist in natural language. Without this distinction it is hard to imagine the possibility of formulating a coherent definition.

Mathematical language is an artificial language based on mathematical logic that leads to a text that can be entirely written in symbols and formulas. The economy of effort implied by such a text is obvious. A few lines from a mathematical text would require dozens of pages written in natural language in order to express the same meanings.

The expression *scientific language*, in its extreme sense, involves a complete mathematical formalization. In reality, the language employed by contemporary sciences (physics, biology, and so on) is a mixture of mathematical language and natural language. It is therefore not surprising that one witnesses endless controversies among scientists about the words and expressions that have an exact meaning a priori. The origin of this semantic morass is elsewhere, particularly in the modern tendency and temptation to regard words as exclusive realities, forgetting (or placing in brackets) the existence of Reality and forgetting that words are only mediators between human beings and Reality. If one yields to this temptation, this can lead to a science that is a pure social construction or to an art or philosophy that is deprived of meaning but acting to undermine social and individual life from within.

A theory of language divorced from a theory of Reality is active nonsense (in the direction of endless growth of illusion). The caducity of a theory of Reality does not imply the slightest need to put Reality itself in brackets, but it does imply the immediate need to change the concept of Reality. In order to achieve mathematical precision, one must understand the precision of Reality.

PEIRCE AND SPONTANEITY

One of the first thinkers of modern times who clarified the intimate relationship between language and Reality is the great American philosopher, logician, and mathematician, Charles Sanders Peirce (1839–1914).

For Peirce, "reality is independent, it does not necessarily belong to thinking in general, but to what you or myself or a limited number of people can think about it."[2] Reality is therefore not separate from thinking but involves the existence of a *direction:* "The activity of thinking that pushes us not towards whatever direction that would depend on our goodwill, but towards a predetermined end, is similar to the operation of fate.... This important law is incorporated into the conception of truth and reality. The opinion that will ultimately have the agreement of all researchers is that what we conceive as 'truth' and the object represented in this view is 'the real.'"[3]

Peirce tried to develop a logic based on graph theory,[4] and thus he arrived very soon, at the age of twenty-eight years, at the conclusion of a *ternary structure* of reality: "a long time ago (1867) I was determined . . . to classify all ideas into three classes of First, Second and Third. . . . No matter how unpleasant it would be to assign significance to numbers and especially to a triad, this thing remains just as true."[5] This ternary structure is the result of Peirce's theory, demonstrated with the help of the graph theory: "any polyad superior to a triad can be analyzed in terms of triads, but a triad cannot generally be analyzed in terms of a dyad."[6] This mathematical result is then linked to the recognition of the "three ways of being": "they are the being of the positive qualitative possibility, the being of the current fact, and the being of the law governing future facts."[7] Finally, the three ways of being are the manifestations of the existence of "three Universes of Experience: the first one contains pure ideas. . . . The second Universe is that of the raw actuality of things and facts. . . . The third universe contains everything that is characterized by the active power of establishing connections between different objects and, in particular, between the objects that exist in different Universes.[8] The three universes are not separate; they "conspire," says Peirce.

Thus Peirce defines the first three basic categories of "Firstness," "Secondness," and "Thirdness": "Firstness is the way of being what it is as it is, positive and without reference to anything else. Secondness is the way of being what it is as it is in relationship to something but without taking into account a third, whatever this may be. Thirdness is the way of being what it is as it is by placing in a mutual relationship a first and a second."[9]

Firstness involves a kind of absolute freedom in regard to representation, a *spontaneity* that inspires and is conscious. "Whatever implies a second is itself a second in relationship with this second one. The first must be present and immediate, so that it is not the second in relationship with a representation. . . . It is also something alive and aware; it has no unity, no fragments. It cannot be thought of in an articulated manner . . . because the assertion always includes the denial of something that is another."[10] The similarity between Peirce's Firstness and the notion of *event* in quantum physics is quite striking.

Firstness implies *a quality of feeling*. "The idea of the present moment," Peirce wrote, "about which, whether it exists or not, we think naturally, as about a point of time in which no thought could take place, in which no detail could be separated, is an idea of Firstness."[11] It corresponds to a potentiality, to a possibility, before existing through an actualization: "The way of being a *redness* before anything was red in the universe, is nevertheless a positive qualitative possibility. And the redness itself, even if it is embodied, is something positive and sui generis."[12]

Peirce perceives Firstness as the dynamic source of development and, at the same time, an element of *indeterminacy* in the formulation of laws: "The only possible way of understanding the laws of nature . . . is to consider them as the result of evolution. This means that they are not absolute, that they cannot be accurately particularized. This implies an element of indeterminacy, of spontaneity . . . in nature."[13] Therefore, there is always "a certain deflection of facts from any well-defined formula."[14] The analogy with the contemporary hypothesis of a Gödelian structure of knowledge is striking. "Things that are indeterminate in themselves have the function to determine each other,"[15] wrote Peirce at the beginning of the twentieth century, in a prophetic presentation of the bootstrap principle, which postulates that a particle is what it is because all other particles exist at the same time. In fact, Peirce had other striking insights that anticipated some fundamental discoveries of twentieth-century science. In 1891, long before the birth of quantum mechanics and superstring theory, Peirce wrote in his famous study "Architecture of Theories": "One should question the validity of the fundamental laws of mechanics for the individual atoms and it seems plausible that they are in a state of motion in more than three dimensions."[16] These insights do not seem accidental. Any thinking based on the ternary structure of Reality is of immediate actuality.

INVARIANCE AND THIRDNESS

If Firstness exposes the monadic aspect of the world, Secondness, in turn, reveals the appearance or existence in space-time by opposition with another, an existence that seems to lack reason: "If I ask you in what the actuality of an event consists, you will say that the actuality consists in the occurrence of the event in that place, at that time," says Peirce. "The actuality of the event seems to lie in its relationship with the universe of existing entities. . . . The actuality has something raw. In it there is no reason."[17] An example of an idea of Secondness is "the experience of the dissociated effort from the idea of a goal to be achieved."[18]

The actuality of an event seems to be something that happens, a response arising from a blind impulse. It becomes coherent when we consider Thirdness as a meaning inherent in *an event of thinking*.

Meaning is the dynamic source of Reality, the very heart of the movement of information: "meanings are inexhaustible. . . . Not only will meaning always shape, for a shorter or longer time, the reactions according to itself, but this is precisely its very being. Therefore, I call this element of the phenomenon or of the object of thought the element of Thirdness . . . the idea of meaning is irreducible to that of quality or creation."[19] So for Peirce, a true triadic relationship involves the participation of thinking. More specifically, the first and the second acquire the qualities of the third, namely of thinking. The first is thinking as a pure possibility, as an *immediate consciousness*. The second is thinking as *information*. Finally, the third is thinking that "determines the idea and gives it form. The third is informational thinking or knowledge."[20] Peirce sees in this authentic Thirdness *the operation of* a sign. The interpreting, mediating thinking between a sign and its object is itself a sign. "The first and the second, the agent and the patient, yes and no, are categories that allow us to broadly describe the facts of experience, and the spirit was satisfied with this for a long time." wrote Peirce. "The third is what throws a bridge over the gap between the first and last absolutes and links them."[21]

Peirce makes a subtle distinction between real and existent. Existent is what appears in opposition with another; the real is what is true, regardless of what people may think about it. So the first, the second, and the third are all real, but only the second exists: "Secondness is the category of existence," wrote Gérard Deledalle.[22]

Peirce calls everything that is introduced to thinking *phaneron*, "but," notes Isabel Stearns, "thinking itself is present within this occurrence."[23] It is as if thinking would know itself. The source of knowledge is thinking nourished by facts of observation. The world of events or *experience* gives a direction to thinking: "only experience teaches us something," wrote Peirce, in a commentary about Roger Bacon.[24] Peirce believed that there are "real things, whose characters are entirely independent of our opinions about them."[25]

Thirdness gives a general character to Secondness; it makes possible the existence of scientific laws through the repetition of facts and the experimental verification of predictions. Predictions cannot be fully realized, because this would require an infinite number of experimental verifications. Thus, *the law implies a certain degree of potentiality*. According to Peirce, there are no exact laws: "exact law obviously never can produce heterogeneity out of homogeneity; and arbitrary heterogeneity is the feature of the universe the most manifest and characteristic."[26] In a sense, *a precise definition implies an inaccuracy of meaning, because it excludes Firstness, the immediate consciousness.* Accuracy of meaning requires the inclusion of Firstness.

Peirce recognizes the fact of being self-corrective as a fundamental characteristic of the scientific method. Here is precisely where the greatness and the limitation of science lie. Invariance is the mirage of science, its daily bread, the base of its success as an operational and rational magic. But this invariance is always approximate, because science cannot fully include Firstness. Its reserved scope is Secondness.

Language, as thinking, exposes the ternary structure of Reality. Specifically, Firstness is always present. In this way, the significance of Lupasco's penetrating statement "a phrase is a real quantum system"[27] can be understood. *The word is the meeting place between continuous and discontinuous, between living and thinking, actualization and potentialization, homogeneity and heterogeneity.* Language is a true quantum phenomenon.

The triadic structure represents the foundation of Peirce's semiotic theory, which is richer and more rigorous than Saussure's dyadic semiologic theory. The relationship established by the sign (whose role is "to make effective ineffective relationships"[28]) is the triad representamen—object—interpretant. The *representamen*, as a visual image or sound of a word, does not possess a completely determined significance. For example, if I pronounce the word "quark," a German will understand that I am speaking about a brand of cheese, an educated man will recognize a word coined by James Joyce, and a physicist will think of constituents of particles. One may need to ask for an interpretant, for example the word *physics*, in order to finally reach the object quark—constituent of particles. The triadic structure led Peirce to a classification of signs according to three trichotomies: "in the first place, according to the fact that the sign itself is a simple quality, a real existent or general law; secondly, according to the fact that the relationship of this sign with its object consists in the fact that the sign has some characters in itself, or in the existential relationship with such or such an object, or in the relationship with its interpretant; thirdly, according to the fact that its interpretant presents it as a sign of the possibility or a sign of the fact or of the reason."[29] Language is thus the border that allows contact between the human being and Reality, the Reality matrix, the life of thinking in its way toward meaning.

THE POSSIBILITY OF A UNIVERSAL LANGUAGE

The concept of evolution can be found at the heart of Peirce's philosophical system. The purpose of ternary dynamics is evolution, the increase of the universe's rationality. One of Peirce's exegetes, Isabel Stearns, writes: "An increasing rationality is not only the property of man, but also of the universe. The thinking should not be isolated, it has its place in nature, and we can find clues about the nature of the world around us in the demeanor of the thinking."[30] The universe is in perpetual movement, but it is a chaotic, anarchic movement; it happens according to a growing *order*, in an evolutionary development dominated by Thirdness. At the social level, the goal of evolution is, according to Peirce, the *universal community*. The universe and, therefore, the human being are becoming more and more rational. What we commonly call "rational" is only/just an approximation of the true rational, and what we call "irrational" is the engine of the rational. In other words, the unknown, the undetermined, is the source of knowledge. There are no fixed, immutable laws of nature waiting to be discovered by us. The very laws of the universe are evolving. This vision is reflected in a study realized by the physicist Walter Thirring on the nature of Nature's laws and is consistent with the vision based on the concept of Reality levels.[31]

Humanity is integrated into this evolution; we must evolve if we want to comply with Reality: "*a trend towards purposes* is in such a necessary manner a feature of the universe *that the action of the hazard on the innumerable atoms* has an inevitable teleological result," Peirce wrote. "One of the aims . . . is the development of the intelligence and knowledge."[32] In the process of evolution, spontaneity—the undetermined Firstness—enables knowledge of Thirdness, the mediator being the world of events, the world of Secondness.

In the dynamics of evolution, we play a privileged role: that of mediator between nature and knowledge. This privileged position of humanity is expressed through the structure of our language: "the word or the sign that the man uses *is* the man himself."[33] The human being is "the most perfect of all signs."[34]

This new language involves the participation of feeling and body. The human being in his or her whole, as an image of Reality, can thus forge a universal language. We do not live only in the world of Secondness, of action and reaction, but also in that of Firstness, of spontaneity and indetermination, and in that of Thirdness, of the thinking that knows itself. The triadic unity of the world must be present in the new language that will link knowledge and being.

Mathematical language and symbolic language foreshadow the new language.

Language, thinking, and symbols cannot be separated: "the life of thought and science is the life inherent in symbols: it is wrong to say simply that a good language is important in order to think well, because language is the very essence of thinking."[35]. Peirce conceives of *the symbol as a living thing*: "in a very strict sense, it is not simply a rhetorical figure; . . . everyone's effort should be to keep intact the essence of each scientific term."[36] It is therefore understandable that an inexact language can have a *destructive effect* on a person: "It is frightening to see how a single unclear idea, a single formula without meaning, hidden in . . . the head, will act at some point as a fraction of inert matter which obstructs an artery, preventing the feeding of the brain."[37]

The health of a human community requires the accuracy of language. The right to the accuracy of language is the right to freedom. With an inaccurate, deliberately obscure, and vague language, beautiful theories can be built and crowds can be attracted, but the result is always the same: illusion and destruction. *Dictatorship through language* is one of the most detrimental ways in which human beings show contempt for other human beings.

Modern mathematical language allows us to address the ineffable precision of reality. At the same time, it is only a *model* of a future universal language, because for the moment it does not include certain key aspects of mythical thinking or of traditional symbolic thinking, which are themselves models of a possible universal language.

The language of Tradition is that of Firstness and the language of science is that of Secondness. The language of Thirdness remains to be found. More and more, an emblematic figure emerges—that of the *polyglot nomad*, following the beautiful expression introduced by Jean-François Malherbe.[38]

CHAPTER TEN

BEYOND DUALISM

> Who rules over contradiction . . . rules over the world.
>
> —Stéphane Lupasco, *Les trois matières*

A STICK ALWAYS HAS TWO ENDS

On the level of theory and scientific experiment, knowledge of the coexistence of the quantum world and the macrophysical world and the development of quantum physics have led to the upheaval of what were formerly considered to be pairs of mutually exclusive contradictories (A and non-A): wave and corpuscle, continuity and discontinuity, separability and nonseparability, local causality and global causality, symmetry and a break in symmetry, reversibility and irreversibility of time, and so forth.

The intellectual scandal provoked by quantum mechanics consists in the fact that the pairs of contradictories it generates are actually mutually contradictory when they are analyzed through the interpretive filter of classical logic. This logic is founded on three axioms:

1. The axiom of identity: A is A.
2. The axiom of noncontradiction: A is not non-A.
3. The axiom of the excluded third: There exists no third term T ("T" from "third") that is at the same time A and non-A.

According to the hypothesis of a single level of Reality, the second and third axioms are obviously equivalent. The dogma of a single level of Reality, arbitrary like all dogma, is so embedded in our consciousness that even professional logicians forget sometimes to mention that these two axioms are in fact distinct and independent of each other.

If one nevertheless accepts this logic—which, after all, has ruled our thinking for two millennia and continues to dominate it today (particularly in political, social, and economic spheres)—one immediately arrives at the conclusion that the pairs of contradictories advanced by quantum physics are mutually exclusive, because one cannot affirm the validity of a thing and its opposite at the same time: A and non-A. The perplexity created by this dilemma is quite understandable: If one is of sound mind, can one assert that night is day, black is white, man is woman, and life is death?

The problem may appear to be merely one of pure abstraction, of interest only to logicians, physicists, or philosophers. In what way is abstract logic important to our daily life?

Logic is the science that has for its object of study the norms of truth or validity. Without norms, there is no order. Without norms, there is no reading of the world; thus, there is no way to apply an idea to our life or our survival. It is therefore clear that a certain logic and even a certain vision of the world is hidden, often unconsciously, behind each action, whatever it is—whether it is the action of an individual, a community, a nation, or a state. The implicit and hidden agenda that determines all social regulation is based in a specific logic.

Since the definitive formation of quantum mechanics around 1930, the founders of the new science have been acutely aware of the problem of formulating a new "quantum logic."

By a fortunate coincidence, this flourishing of quantum logics coincided with the flourishing of a new mathematically rigorous formal logic, which tried to enlarge the range of classical logic. This phenomenon is relatively new because for two millennia, human beings had believed that logic was unique, stable, given once and for all, inherent in their own brains.

However, there is a direct relation between logic and the environment—physical, chemical, biological, macro-, and micro-sociological. Like knowledge and comprehension, the environment changes with time. Therefore, logic can only have an empirical foundation. The notion of the "scientific" or "mathematical" logic is very recent—it appeared in the mid-nineteenth century. Another major idea—that of the history of the universe—appeared shortly afterwards. Previously, the universe, like logic, had been considered to be eternal and unchangeable.

History will credit Stéphane Lupasco with having shown that the logic of the included third is a true logic, formalized, multivalent (with three values: A, non-A, and T) and noncontradictory. Like Edmund Husserl, Lupasco was a pioneer. His philosophy, which takes quantum physics as its point of departure, has been marginalized by physicists and philosophers. Curiously, it has also had a powerful, albeit underground, impact among psychologists, sociologists, artists, poets, theologians, and historians of religion. Lupasco was right too soon. Perhaps the absence of the notion of "levels of Reality" in his philosophy obscured the substance of his philosophy. Many persons believed that Lupasco's logic violated the principle of noncontradiction—hence the rather unfortunate name "logic of contradiction"—and that it entailed the risk of endless semantic glosses. The visceral fear of introducing the idea of the included third, with its magical resonances, only helped to increase the distrust of such a logic.

Our understanding of the axiom of the included third—that there exists a third term, T, which is at the same time A and non-A—is completely clarified once the notion of "levels of Reality" is introduced.

In order to obtain a clear image of the meaning of the included third, we can represent three terms of the new logic—A, non-A, and T—and the dynamics associated with them by a triangle in which one of the vertices is situated at one level of Reality and the two other vertices at another level of Reality. The included third really is an *included third*. If one remains at a single level of Reality, all manifestation appears as a struggle between two contradictory elements (example: wave A and corpuscle non-A). The third dynamic, that of the T state, is exercised at another level of Reality, where that which appears to be disunited (wave or corpuscle) is in fact united (quanton), and that which appears contradictory is perceived as noncontradictory.

It is the projection of the T state onto the same single level of Reality that produces the appearance of mutually exclusive, antagonistic pairs (A and non-A). A single level of Reality can only create antagonistic oppositions. It is inherently self-destructive if it is completely separated from all the other levels of Reality. A third term, let us call it T_0, which is situated at the same level of Reality as that of the opposites A and non-A, cannot accomplish their reconciliation. The "synthesis" between A and non-A is above all an explosion of immense energy, like that produced by the encounter between matter and antimatter. In the hands of Marxist-Leninists, the Hegelian synthesis appeared like the grand finale of progressive development on the historical plane: primitive society (thesis)—capitalist society (antithesis)—communist society (synthesis). Alas, it has metamorphosed into its opposite. The unexpected fall of the Soviet empire was in fact inexorably inscribed in the binary logic of its own system. Logic is never innocent. It can even cause millions of deaths.

The entire difference between a triad of the included third and a Marxist triad is clarified by a consideration of the role of time. In a triad of the included third, the three terms coexist at the same moment in time. This is why the Marxist triad is incapable of accomplishing the reconciliation of opposites, whereas the triad of the included third is capable of it. In the logic of the included third, the opposites are rather contradictories: the tension between contradictories builds a unity that includes and goes beyond the sum of the two terms.

One also sees the great dangers of misunderstanding engendered by the common-enough confusion between the axiom of the excluded third and the axiom of noncontradiction. The logic of the included third is noncontradictory in the sense that the axiom of noncontradiction is thoroughly respected: it is a condition that enlarges the notions of "true" and "false" in such a way that the rules of logical implication concern not just two terms (A and non-A) but three terms (A, non-A, and T), all coexisting at the same moment in time. This is a formal logic, just as any other formal logic: its rules are derived by means of a relatively simple mathematical formalism.

One can see why the logic of the included third is not simply a metaphor, like some kind of arbitrary ornament for classical logic, which would permit adventurous incursions and passages into the domain of complexity. The logic of the included third is perhaps the privileged logic of complexity—privileged in the sense that it allows us to cross the different areas of knowledge in a coherent way.

The logic of the included third does not abolish the logic of the excluded third: it only constrains its sphere of validity. The logic of the excluded third is certainly valid for relatively simple situations; for example, driving a car on a highway: no one would dream of introducing an included third in regard to what is permitted and what is prohibited in such circumstances. On the contrary, the logic of the excluded third is harmful in complex cases, for example, within the social or political spheres. In such cases it operates like a genuine logic of exclusion: good or evil, right or left, women or men, rich or poor, whites or blacks. It would be revealing to undertake an analysis of xenophobia, racism, anti-Semitism, or nationalism in the light of the logic of the excluded third.

The ancient Chinese saying that "a stick always has two ends" expresses something very profound. Let us imagine, as in the story "The End of the End," told by the famous French artist Raymond Devos, a man who desperately wants to separate the two ends of a stick. He cuts his stick and then sees that instead of having separated the two ends, he now has two sticks, both of which have two ends of their own. He goes on cutting his stick, all the while becoming more and more anxious—the sticks multiply ad infinitum, but he finds it impossible to separate the two ends!

Are we, in our present civilization, in the same position as this man who applies himself so assiduously to trying to separate the two ends of his stick? Only the intelligence of an included third can answer the barbarism of the excluded third, because a stick always has two ends.

STÉPHANE LUPASCO (1900–1988): THE HERALD OF THE COMING THIRD

The first attempts to formulate a quantum view of the world were made at the edge of the contemporary philosophical movement, thanks to the work of a physicist, Niels Bohr (1885–1962); an engineer, Alfred Korzybski (1879–1950); and a philosopher of scientific formation, Stéphane Lupasco (1900–1988). Therefore, in the first half of the last century, we can find the emergence of three main directions:

That of Bohr, who was convinced that the principle of complementarity could be the starting point of a new epistemology and embrace physics, biology, psychology, history, politics, and sociology at the same time[1]

That of Korzybski, who proposed a non-Aristotelian system of thought, with an infinity of values[2]

That of Lupasco, based on the logic of energetic antagonism

In this context, the works of Stéphane Lupasco have a special place. The principle of complementarity represented a very restricted basis, and Korzybski's approach, despite the important contributions to the understanding of language structures, remained too vague. Lupasco was the only one capable of identifying a law of invariance, which allowed, in principle, the unification of different fields of knowledge.

Although he wrote fifteen books, and although artists, scientists, thinkers, and intellectuals of high quality such as Gaston Bachelard, Gilbert Durand, Edgar Morin, Henri Michaux, André Breton, Salvador Dali, Georges Mathieu, René Huyghe, Thierry Magnin and André de Peretti admitted the importance of his work, Lupasco remains a little-known philosopher. A major international congress, "Stéphane Lupasco: The Man and the Work," took place in March 1998 at the Institut de France. Following this congress, a collective volume was issued that highlights the verity of Lupasco's philosophy.[3]

The belief that the most general results of science must be integrated into any philosophical approach crosses the entire work of Lupasco like an axis: "no theory, no doctrine, no outlook . . . is really possible any longer if it ignores the data of the scientific experiment, which inundates everything;

on the other hand, we can no longer nurture ourselves with the theoretical acquisitions of the existing body of knowledge, because they no longer answer to our needs.[4]

One of the best illustrations of Lupasco's logic of antagonism is provided by the gradual historical evolution of his own philosophical thinking. This thinking is placed under the double sign of discontinuity with the existing philosophical thinking and of continuity (hidden, because inherent to the very structure of the human thinking) with Tradition. The principle of dualistic antagonism has been formulated fully since 1935, in his thesis *Du devenir logique et de l'affectivité*.[5] The book opens with a profound meditation on the contradictory nature of space and time revealed by Einstein's restrained relativity theory, which is the apogée of classical physics. The notions of actualization and potentialization are already present in this meditation, even if they will only be gradually specified later in his books at the level of the reader's current understanding, but also at the level of correct terminology.

A second step was made with the issue of *L'expérience microphysique et la pensée humaine*, published in 1941.[6] In this book, Lupasco assimilates and extends the teaching of quantum physics in a veritable quantum view of the world.

According to Lupasco's interpretation of Heisenberg's relations, the actualization of spatial localization triggers the potentialization of the quantity of movement, and the actualization of temporal localization triggers the potentialization of extension in energy. In this way, Lupasco integrates into these ideas the fact that the concept of the identity of a particle, in the classical meaning of the term, is no longer valid in the quantum world.

Lupasco recognized the universal dimension of Planck's discovery in its entire scope: "Of course, there is no problem that is more enigmatic than that of the emergence of quanta. How did Planck have this awesome idea of quantification? This is a psychological and historical event that has its origin . . . in the most remote metaphysical movements of human thought and destiny." He sees in the sudden emergence of the discontinuity the sign that announces a change in history: "Planck's intuition . . . is similar to some of those short, modest, and inexplicable historical acts that change the course of human events for a long period." Thus, Lupasco feels entitled to formulate the crucial problem, that of extrapolating a scientific idea to Reality as a whole: "Should quantification, that is, in our opinion, precisely the irresistible—and unconscious—introduction of contradiction in the midst of microphysical facts . . . be extended to all facts?"[7] In the same book, Lupasco highlights the philosophical importance of Pauli's exclusion principle, a true principle of individuation in the evanescent world of particles.

The last decisive step was made in 1951, with *Le principe d'antagonisme et la logique de l'énergie*, which represents Lupasco's attempt at an axiomatic formalization of the logic of antagonism.[8] This formalization is important for the crystallization of Lupasco's thinking, because it introduces a rigor, an accuracy, without which this thinking could only be considered a huge, fascinating, but vague reverie.

Lupasco's philosophy has as a starting point modern physics and axiomatic logic. The most general results of science can and should be integrated into the very foundations of a philosophical approach, provided that nature is not an accident of existence. In this respect, Lupasco's philosophy is of great novelty. This philosophy is based on science and then returns to science in order to fertilize it, to ennoble it with a unified view of the world that can only hasten important scientific discoveries.

One major objection can be made immediately, however: how can philosophy, in its desire for stability and permanence, accept science as a foundation, knowing that science is in a state of perpetual seething, of continuous change? It is true that Lupasco's philosophy is based on the broader results of contemporary science, but it tries to extract from these results what is even more general in a search for invariance and universality. In fact, it is precisely in this *invariance*, in this search for *general laws* that cross all scales and that govern phenomena at all scales, that the intimate bond between Lupasco's philosophy and Tradition consists. According to Lupasco, *invariance is the logic of energy*.

THE INCLUDED THIRD

The included third does not mean at all that one could affirm one thing and its opposite—that, by mutual annihilation, would destroy any possibility of prediction and therefore any possibility of a scientific approach to the world.

It is rather about recognizing that in a world of irreducible interconnections (such as the quantum world), conducting an experiment and interpreting experimental results are inevitably reduced to a cutting up of the real, which affects the real itself. The real entity can thus reveal some contradictory aspects that are incomprehensible, even absurd from a logical point of view, based on the assumption "*either* this one *or* the other one." These contradictory aspects cease to be absurd in a logic based on the assumption "*both* this one *and* the other one" or rather, "*neither* this one *nor* the other one."

Due to the rigorous development of his axiomatic formalism, Lupasco postulated the existence of a third type of antagonistic dynamics, which coexists with that of *heterogeneity*, which governs living matter, and with that of *homogeneity*, which governs macroscopic physical matter. This new dynamic mechanism assumes the existence of a state of rigorous, precise balance between the poles of a contradiction, in a strictly equal quasiactualization and quasipotentialization. This state, which Lupasco calls the *T state* ("T" being the initial for the included "third"), characterizes the quantum world and the psyche world.

In 1936, Birkhoff and von Neumann presented the first proposal of such quantum *logic*. Since then, numerous works (those of Mackey, Jauch, Piron, and others) have been devoted to the study of a coherent formulation of quantum logic.[9] The ambition of such logic was to solve the paradoxes generated by quantum mechanics and to attempt, wherever possible, to reach a higher predictive power than that reachable through classical logic. Its status remains ambiguous: one doubts its predictive power and even its existence as a general theory of valid inferences. Most quantum logic changes the second axiom of classical logic—the axiom of noncontradiction—by introducing noncontradiction with several truth values instead of with the binary pair (A, non-A). Such multivalent logic fails to consider another possibility: the modification of the third axiom—the axiom of the excluded third. It cannot be said that Lupasco's axiomatic formalism is per se quantum logic in the sense that it could be directly applied to the specific, detailed inferences of quantum mechanics. This formalism should first be translated according to the terminology of quantum physics.

THE TERNARY DIALECTICS OF REALITY

According to Lupasco's logic, energy, "in its fundamental constituents, owns at the same time the property of identity and the property of the distinguishing differentiation."[10] The manifestation of a certain phenomenon is equivalent to a certain *actualization*, to a trend toward identity, but this manifestation itself involves a certain repression, a *potentialization* of everything that this phenomenon is not; in other words, of non-identity. Potentialization is not annihilation, disappearance, but rather memorization of *what-has-not-yet-been-manifested*. This concept is a direct translation of the quantum situation.

An immediate consequence of the introduction of the concept of potentialization is the fact that this local causality (that of actualization) is always associated, in Lupasco's approach, with an antagonistic finality. Local causality only appears in a narrow field of Reality. Global causality is present at all levels of Reality. Reality as a whole is nothing but a perpetual oscillation between actualization and potentialization. Taking into account only actualization inexorably leads to deformed Reality. There is no absolute actualization.

But actualization and potentialization are not sufficient for a coherent logical definition of Reality. Movement, transition, going from the potential to what is manifested, is not conceivable without an *independent dynamism* that involves a perfect, rigorous balance between actualization and potentialization, a balance that makes this transition possible.

It is interesting to mention that the notion of "potentialization" appears almost simultaneously in Lupasco's work and Heisenberg's work. Moreover, Heisenberg is fully aware that only a logic of the included middle could help us understand the quantum paradoxes, as the great Romanian writer and Goncourt Prize winner Vintila Horia reports in a book describing his meeting with Heisenberg.[11] Heisenberg completely ignored the work of Lupasco; however, Lupasco nurtured a genuine admiration toward Heisenberg and his uncertainty principle but made no effort to contact him. The great physical meeting between Heisenberg and Lupasco did not occur. That was the will of destiny.

A different perspective on the ternary structure of Lupasco's philosophy can be obtained by analyzing the concepts of *homogeneity* and *heterogeneity* that he introduced. Homogeneity is the process of moving toward sameness, toward an endless accumulation of all systems in one and the same state, toward total disarray, toward death (conceived of as lack of movement). The physical origin of this concept is the second law of thermodynamics (or Carnot-Clausius's principle), which shows that for a closed macrophysical system, entropy increases, chaos increases, and energy degrades toward heat. In the microphysical world, homogeneity governs the evolution of particles, such as photons, which are not subject to Pauli's exclusion principle: they can accumulate indefinitely in the same quantum state. "Thus, the universe dies in the light," wrote Lupasco. For movement to be possible, the homogeneous and the heterogeneous must coexist. *The homogeneity-heterogeneity antagonism is thus an organizing, structuring dynamism.*

Again, the antagonist couple heterogeneity-homogeneity is not sufficient to ensure movement. A third dynamism is required that involves the right, rigorous balance between homogeneous and heterogeneous (and which is, therefore, neither homogeneous nor heterogeneous).

Lupasco's axiomatic logic thus releases three preferred orientations, three dialectics: *the dialectic of homogeneity*, *the dialectic of heterogeneity*, and *the quantum dialectic*. Lupasco employs the term *tridialectics* to characterize the structure of his philosophical thinking, a term that expresses the ternary, tripolar structure (homogeneous-heterogeneous-T state) of any manifestation of Reality, the *coexistence* of these three inseparable aspects inside any dynamism that is available to logical, rational knowledge.

With its origin in quantum physics, Lupasco's tridialectics represents a general reading grid for a wide variety of phenomena. Thus the presence of the axiom of the included third causes a relationship between Lupasco's approach and the *symbolic thinking* that consists in having many common consequences. Facts that are as far from quantum physics as ethnographic or anthropological ones thus find a possibility of consistent interpretation in Lupasco's philosophy. Gilbert Durand's testimony is significant in this regard: "our empirical research reached a level of the classification of images, all governed by three principles, and . . . Stéphane Lupasco, without going through the mediation of anthropological or ethnographic investigation, established a system of logic with . . . three terms that roughly coincided with the 'three logics' that Roger Bastide and I have found in our anthropological research. Thus, the concrete coherence of symbols (isotopism) in the midst of the constellations of images also revealed this dynamic system of antagonistic 'cohesive forces,' for which the logics only represent the formalization."[12]

TRIADIC SYSTEMOGENESIS AND THE THREE MATTERS

The antagonistic dynamisms, in their various balances, create *systems*. These systems represent the structuring of matter, the perception through the sense organs being just an appearance, an illusion.

The energetic antagonism involves an undefined chain of contradictories: "two antagonistic dynamisms giving rise to a system, this system . . . will involve an antagonist system of the same order; these two systems will involve a system of antagonistic systems, and so on, according to what we named *systemogenesis*," wrote Lupasco.[13]

The ternary structure of energetic systematizations is reflected in the structuring of three types of matter, which are not isolated or separated: "matter is not based on the 'dead' . . . in order to rise through the biological, from complexity to complexity, to the psyche and beyond: the three aspects represent . . . three divergent orientations, of which one, of microphysical nature . . . is not a synthesis of the other two, but rather their struggle, their inhibitory conflict."[14] The conclusion that *any event, any system involves a triple aspect—macrophysical, biological, and quantum (microphysical or mental)*—is certainly surprising and full of many consequences.

Lupasco never asserted that the microphysical world *is* the psyche world. He only highlighted the *isomorphism* between these two worlds that results from the fact that there is nothing of an absolute actuality (reality) and nothing of an absolute potentiality. Lupasco never said that the soul is located inside the electron or the proton or the muon or the neutrino; such a claim would have been, indeed, absurd, because each of the hundreds of known particles is as fundamental as all the other known particles. The microscopic world and the psyche world are two *different* manifestations of a single tridialectic dynamism. Their isomorphism, similar to that studied by Pauli and Jung, is ensured by the continuous, irreducible presence of the T state in any event.

NONSEPARABILITY AND THE UNITY OF THE WORLD

At a superficial glance, one might think that the antagonism, the struggle between the contradictories, involves their separation. In fact, it is the opposite. Lupasco's formal logic ineluctably induces *nonseparability*: "there is no element, no event, no arbitrary point in the world that is independent, that is not in a specific relationship of connection or breakage with another element or event or point, as there is more than one item or event or point in the world. . . . Therefore, everything is linked in the world."[15] This statement is not a postulate, but a *result* of Lupasco's logic. It is even intuitively comprehensible: because the existence of any one system means the existence of an antagonist system, it results that any two systems will be linked by a chain of antagonistic systems. *Energetic antagonism is thus a view of the world's unity*, a dynamic unity, a unity of an indefinite chain of contradictories based on a universal ternary structure.

THE NATURE OF SPACE-TIME

One of the most interesting aspects of Lupasco's philosophy is the presence of the T state. The rigorous balance between actualization and potentialization, which causes a maximal densification of energy, seems to indicate that no direct manifestation of this state in our continuous space-time is possible: the space-time associated with the T state is of a different kind from continuous space-time. But any energetic event has a ternary structure. The T state must necessarily coexist with event states, both those with a heterogenizing and those with a homogenizing tendency. Thus we reach the apparently paradoxical conclusion that continuous space-time is not enough to describe reality: a broader space must be defined that would include continuous space-time in one way or another. Local causality defined in continuous space-time is no longer valid in this broader space. Continuous time appears as an approximation. *Is this T state the origin of quantum indetermination, of discontinuity, of nonseparability, of nonlocalization, of the extra dimensions of the M theory?*

The theory developed by the mathematician Roger Penrose[16] tosses out the hypothesis of the space-time continuum, a hypothesis that, according to Penrose, has no real physical evidence. The theory developed by T. D.

Lee, Nobel laureate in physics, is based on the same idea. The M-theory of the superunification of physical interactions goes even further: space-time is no longer a fundamental concept.

At first, as a result of his logic, Lupasco noted the primacy of the relationship over the object. Consequently, time and space are themselves a *result* of the logic of contradictory antagonism.

Time is the result of movement, of change, of logical dynamism. *Time is therefore born from the conflict between identity and diversity.* Space is also a result of logical dynamism. Space appears as a *contradictional conjunction*; time appears as a *contradictional disjunction*: space and time are linked by a relationship of contradiction. In Lupasco's words, "There will always be space in time and time in space."[17]

Actualization and potentialization no longer occur in space-time, but space-time is generated by the contradiction actualization-potentialization.

Finally, each pole of the ternary structure homogeneous-heterogeneous-T state leads to a proper space-time. Therefore, *the time corresponding to an actualization will necessarily be discontinuous*, as it follows from the simultaneous action of the three poles with their associated space-times: "Any time evolves jerkily, by leaps, by advances and retreats, because of the very constitution of the dialectics which gives rise to it . . . Logical temporality is thus discontinuous."[18]

The fecundity of Lupasco's approach in contemporary studies in music,[19] literature,[20] politics,[21] ethics,[22] epistemology,[23] psychology,[24] linguistics,[25] and art theory[26] is breathtaking. This was evoked in a recent UNESCO international symposium held in 2010 in Paris.[27] Also, the consequences of Lupasco's included third for the dialogue between science and religion have been explored, and these investigations are particularly inciting.[28]

IS LUPASCO A PROPHET OF THE IRRATIONAL?

Reason is contradictory in its own nature.[29]

Where does the rational end and where does the irrational begin? Gilles Gaston Granger[30] rightly distinguishes three types of irrational: the irrational "as an obstacle, as a starting point for a conquest of rationality," "the irrational as an appeal, as a means to renew and extend the creative act," and finally, the irrational "by cancellation," which corresponds to "a genuine rejection of the rational." The whole history of science testifies to the continuous and bitter fight against the unknown, and it could be asserted that the unknown is the very source of scientific advance. Many issues that were once considered irrational, bizarre, and paradoxical became, due to the scientific approach, rational, normal, and integrated into a coherent scientific description. From this point to asserting that everything in the world is rational is a big step, of which the consequences, at all levels, should not be underestimated.

The materialist-dialectic view of the world tells us that everything that is unknown in the world will one day be revealed, known. The irrational is thus designed as an *asymptotic*, abstract point, lacking any real character. What is real is rational; therefore it must be subjected to reproducibility, testing, and scientific understanding.

We can view this materialist-dialectic description of knowledge, representing scientific knowledge, as the Nobel laureate in Physics David Gross suggests (even if he does not even remotely declare himself, at least not explicitly, an advocate of the materialist-dialectic approach), as a sphere.[31] The *surface* of the sphere is the boundary between the known and the unknown. The sphere is *compact:* all the points inside the sphere represent what is known at a certain point in time. In ongoing efforts throughout history, humanity keeps pushing the frontier of the sphere ever further, in an endless process. In this process of scientific knowledge, the volume of the sphere (what is known) increases and, simultaneously, the surface of the sphere (the border between the known and the unknown) increases. But the volume of the sphere is growing faster than its surface: The *ratio* between volume and surface tends toward infinity with the passage of time, thus ensuring the continuous and endless advance of scientific rationality. This image, finer and more subtle than that which is traditionally proposed by dialectic materialism, leads to the same conclusion: the irrational has no value in reality.

It is possible to formulate a radically different view, based on the quantum and systemic views of the world, despite superficial analogies. Let us consider again a sphere as the representation of scientific knowledge, but a *non-compact* sphere: within the sphere of the known there are small spheres representing the unknown. In the process of scientific knowledge, the small spheres *decrease* both in volume and surface, the ratio between volume and surface tending, with time, toward *zero*. Therefore, in this respect also, there is a progression in scientific knowledge. But the unknown is constantly present, in an irreducible manner: it is manifested by points that will be present, whatever we do, inside the sphere of the known. And could not we define *the sacred* as being precisely all that is irreducible in relation to mental operations? Everything happens as if there was no opposition but a permanent *cooperation* between rational and irrational, which manifest as two contradictory poles of the same Reality, which transcends both of them. Everything happens as if there was a mutual interaction, a mutual transformation between rational and irrational. The irrational does not appear as the attribute of an entity outside the scope of the knowable sphere but as the pole of a dynamism that includes it and at the center of which the human being is placed. This dynamism is a source of freedom, of spontaneity, and of creativity in the evolution of natural systems.

This picture is quite similar to the conclusions that emerge from Edgar Morin's work.[32] In his book *Science avec conscience*, he stresses the need for a new way of thinking: "to think does not mean to serve order or disorder, it means to make use of order and disorder. To think does not mean to move apart from the irrationalizable and the inconceivable. It means to work in spite of/against/with the irrationalizable and the inconceivable." Morin recommends the adoption of open thinking, accepting negotiation with the unknown, and of a knowledge that is aware of the ignorance that it brings. This open thinking must necessarily be founded on a *new rationality*: "A new rationality is looming. The old rationality was only fishing. But it caught no fish, only fish bones! By allowing us to conceive of conceptualization and existence, the new rationality allows us to perceive not only the fish, but the ocean as well, which is what could not be caught."[33]

Finally, this image is also close to Max Planck's conception of the role of the irrational in scientific knowledge: "physics, like any other science, contains a kernel of irrationality that is impossible to fully reduce. And yet, to consider this irrational—as being outside science by definition—would mean to deprive science of all its internal dynamism. The cause of this irrationality, as modern physics shows ever more clearly, is that the scientist himself is one of the constituent parts of the universe."[34]

THE EXPERIENCED THIRD

The ternary structure of Reality is present inside the human being: the intellectual center represents the dynamism of the heterogeneity; the motor center, the dynamism of the homogeneity; and the feeling center, the dynamism of the T state. A person's whole life is an oscillation between the three poles of the ternary. The intellectual dynamism (through its mentally truncated shape) can lead to death by extreme differentiation (and this may happen if science becomes the unique, absolute religion of the person). The motor center dynamism can lead to death by the achievement of an absolute, noncontradictory identity (and this can happen if wealth and material comfort are the human's only concern). The feeling dynamism thus appears as a protector of life. Recent findings in neurophysiology clearly show the exceptional role of feelings in a human being's life.[35] In particular, a viable education can only be an integral education of the human being. It is founded on a new type of intelligence requiring harmony between mind, feelings, and body.

Totalitarian societies, with their homogenizing tendency, are built on a belief in absolute actualization, on the willingness to transform the contradictories into opposites. These societies do not know that they are doomed to death from the very beginning. On the other hand, democratic societies are also founded on a belief in absolute actualization: that of heterogeneity. Despite their considerable differences, democratic societies have a fundamental characteristic in common: *that of the progressive potentialization of the T state*. The world will know, one day, a new, tridialectic society, based on the progressive actualization of the T state, which involves a rigorous balance between homogeneity and heterogeneity, between socialization and maximal individual fulfillment.

Wars are based on the same fanatical dedication to absolute actualization. The imbalance of the ternary, the preferential development of one or another direction through the suppression of the contradiction, equates, according to Lupasco's logic and philosophy, to a redoubtable pathology. Wars are, in this respect, demonstrations of immense collective psychosis.

Thus the philosophy of the included third appears as a philosophy of *freedom* and *tolerance*. In a very stimulating study, Jean-François Malherbe showed how interaction between the included third and Wittgenstein's language games could have important repercussions in regard to the formulation of a contemporary ethic.[36] As with any philosophy that deserves the name, for it to be operative, the included third must be experienced and applied in everyday life.

CHAPTER ELEVEN

THE PSYCHOPHYSICAL PROBLEM

REDUCTION AND REDUCTIONISM

The concept of the psychophysical (the link between the physical, biological, psyche, and spiritual levels of the human being) acutely raises into question its conflict with reductionism (in which the physical, biological, psyche, and spiritual are considered to be one single level of Reality). In order to clarify this issue, it is first of all necessary to distinguish two words that are very central in modernity: *reduction* and *reductionism*.

The scientific meaning of the word *reduction* is the following: A is reduced to B, B to C, C to D, and so on until we reach what is considered to be the most fundamental level. Indeed, human thinking aims for the same process of reduction. Reduction is, in many ways, a natural process of thinking, and there is nothing wrong with that. The only question concerns the understanding of what is at the end of the reduction chain: is this chain circular? If not, how can we justify the concept of the *end* of a chain?

With regard to *scientific reductionism*, things are very different. This designates the explanation of spiritual processes in terms of mental processes, which in turn are explained in terms of biological processes, which in turn are explained in terms of physical processes. In other words, scientists who comply with the life of their community reduce spirituality to materiality.

Philosophical reductionism reverses the chain: it reduces materiality to spirituality. The two types of reductionism belong to what might be called *monoreductionism*. Some philosophers adopt a dualistic approach: they consider that materiality and spirituality are fundamentally distinct. The dualistic approach is a version of philosophical reductionism: it corresponds to what we may call *multireductionism*. We can even identify a different version in New Age literature: that of *interreductionism*—certain properties of material nature are attributed to spiritual entities, or, conversely, certain properties of spiritual nature are attributed to material objects.

Antireductionism, the approach that is opposed to reductionism, is expressed through *holism* (meaning that the whole is more than the sum of its parts and determines the properties of its parts) and through *emergentism* (which means that new structures, behaviors, and properties are generated by relatively simple interactions that give rise, in turn, to increasing levels of complexity). Holism and emergentism have their own difficulties: they must explain where *novelty* comes from without providing ad hoc arguments.

The concept of *levels of reality* is crucial in order to reconcile reductionism, which is so necessary to the scientific approach, with antireductionism, which is so necessary to the study of complex systems.

THE *COINCIDENTIA OPPOSITORUM* AND HERMETIC IRRATIONALISM

Speaking about the revival of hermetic thinking, Umberto Eco says:

> Hermetic knowledge influenced Bacon, Copernicus, Kepler, and Newton, and modern quantitative science will be conceived, among other things, from an exchange with that qualitative knowledge of hermetism. . . . However, this influence is intimately linked to a certainty that hermetism did not have, of which it could not and did not want to be aware: the description of the world was made according to a logic of quantity and not according to one of quality. Thus, paradoxically, the hermetic model contributes to the birth of its new enemy: modern scientific rationalism. Hermetic irrationalism will then migrate to mystics and alchemists, on the one hand, and to poets and philosophers, on the other hand: from Goethe to Nerval and Yeats, from Schelling to von Baader, from Heidegger to Jung.[1]

A little farther on, Eco refers directly to Jung: "When he revisits the old hermetic doctrines, Jung brings again into question the gnostic issue of the rediscovery of an originating self."[2] Finally, Umberto Eco describes for us what the *hermetic drift* consists in:

> The main feature of the hermetic drift seemed to be the uncontrolled ability to slip from significance to significance, from resemblance to resemblance, from one connection to the next. Contrary to contemporary theories of the drift, hermetic semiosis does not assert the absence of a universal, univocal, and transcendental signified. It assumes that everything—excluding the mysterious connection

in rhetoric—can refer to anything else, precisely because there is a strong transcendent subject, the neoplatonic One. This—being the principle of the universal contradiction, the place of *coincidentia oppositorum*, alien in regard to any possible determination and therefore at the same time Everything, Nothing, and Inexpressible Source of Everything—acts in such a manner that everything connects to everything else due to a labyrinth-spiderweb of cross-references. Thus it seems that hermetic semiosis identifies in any text (for example, in the Great Text of the World) as the Fullness of the Signified, not its absence. And yet, this world, invaded by signatures and governed by the principle of universal significance, gave rise to effects of constant sliding and rejection of any possible signified.[3]

Umberto Eco's diagnosis is seductive, but it is based on both a logical and an epistemological error. There is certainly a *confusion*, which translates into an unacceptable mixture of levels of understanding. The level of understanding of Jung and Lupasco's thinking is radically different from that (conceived in one reduction or another) of parapsychology, the New Age, or contemporary hermetic irrationalism. How was it possible for such a huge confusion to occur?

Among the three trends mentioned, the confusion with hermetic irrationalism is the hardest to identify, and therefore it is the most redoubtable. After all, the confusion regarding parapsychology can be explained by the undeniable attraction that Jung, Pauli, and Lupasco felt toward so-called "paranormal" phenomena. Moreover, the confusion about the New Age can be explained by the intellectual neglect of those who are engaged in this movement. But at the end of the day, is not assimilation with hermetic irrationalism legitimate?

THE CORE OF THE PROBLEM: WE ARE TOO DEEPLY IMMERSED IN THE SEVENTEENTH CENTURY

The difficulty of an answer is compounded by the fact that we are faced with some real mutants. Carl Gustav Jung (1875–1961) is a mutant, necessarily solitary in his field, who belongs to the soft sciences. Wolfgang Pauli (1900–1958) is also a mutant in his field, as he belongs to the hard sciences. Lupasco, is, of course himself a mutant, with his logic of the included third. It is true that Pauli is a less-solitary mutant since physics suffered an unprecedented *collective* mutation through the emergence of quantum mechanics.

But all three are all confronted with the same problem: the *nonconformity between studied phenomena and the model of Reality* that prevailed in their times. To this problem, which must certainly have been felt as dramatic on a personal level, there were two possible solutions.

The first was to circumvent the problem simply by placing reality between brackets, as an obscure, ontological, and unnecessary concept for scientific progress. This is the way that has been followed by most scientists.

Jung, Pauli, and Lupasco decided on a far more difficult and painful solution—that of the invention (or perhaps the discovery) of a new model of Reality. "When a regular person says, 'reality,' that person thinks of it as something obvious and well known," says Pauli. "But, for me, the formulation of a new idea of reality is the most important and most difficult task of our time . . . What I have in mind—provisionally—*is the idea of the reality of the symbol*. On the one hand, a symbol is the result of a human effort and, on the other hand, it indicates the existence of an objective order in the cosmos, of which humans are only a part."[4]

The meeting between Jung and Pauli was predestined by the common character of their philosophical and existential concerns. And this meeting was updated, as by chance, by a disease: following a profound depression that had its onset in the winter of 1931, Pauli became Jung's patient. It was a happy event that would generate an interaction of considerable importance for our age. Certainly, as is natural, the interaction worked both ways: the thinking of a great physicist influenced, in an undeniable way—especially in the understanding of the principle of complementarity of quantum mechanics and the idea of causal order—the thinking of a great psychologist, and the great psychologist softened the thinking of the great physicist. Pauli testified several times about the influence of Jung's thinking on his own thinking. For example, in a letter addressed to Markus Fierz, professor of theoretical physics at the University of Basel, he wrote, with regard to his article "Fundamental physics": "I see my essay as a modification and expansion of Bohr's analogies about physics and psychology, due to *the acceptance of the concept of 'unconscious'* as it is used by the modern psychologists of the Jungian school and of other schools."[5]

However, this mutual interaction does not mean their points of view became the same. For example, Pauli wrote several times that he did not understand the meaning that Jung attributed to the phenomena of synchronicity and even claimed that Jung's astrological study was a total failure.[6] But at the same time, Pauli contributed substantially to the development of the concept of synchronicity.

It is legitimate to question the possibly salient role played by the depressive experience in the development of Jung and Pauli's thinking. The experience of depression opened the valves of the imaginary and of feelings. However, these depressive experiences were asymmetric.

In Jung's case, the experience had as a background the *emptiness* of the intellect; thinking was as if blocked, suspended at the gates of the unconscious realm. One of the most vivid traces of this experience is the superb work *The Seven Sermons for the Dead*.[7] Jung's testimony is very valuable and unambiguous: "During the years when I have listened, my inner images represented the most important period of my life, during which all essential things were decided. It is then that they started, and the details that followed were only complements, illustrations, and clarifications. Further on, my whole work was to elaborate what had been rushing out from my unconscious during those years and that overwhelmed me at first. It was the raw material for the work of a lifetime."[8]

For Pauli, it was rather about a fullness of the intellect. Pauli was a gifted child. At age nineteen, he was already the author of an encyclopedic volume on the theory of relativity, a volume praised by Einstein himself. By the age of twenty-four, he had made his major scientific discovery—the principle of exclusion—which bears his name and would bring him the Nobel Prize in 1945. The best proof of that experience is his conviction that *the most important issue of our time is the psychophysical issue*—the relationship between spirit and matter—an issue that he would continue to probe more deeply up to the premature end of his life. For Pauli, the holistic nature of *unus mundus* implied that the states describing the material and psychological fields are nested. He believed that future science would conceive of reality as being simultaneously physical and mental, although it goes beyond the physical and the mental.[9]

The written philosophical work of Pauli has been reduced to a few articles, all of them in German. His work is instead formulated in his correspondence, which reveals a Pauli who was unknown to the general public—a great thinker, certainly one of the greatest among the founders of quantum mechanics. The Finnish physicist Kalervo Laurikainen made available to the public a few fragments of Pauli's correspondence with Markus Fierz in his book *Beyond the Atom: the Philosophical Thought of Wolfgang Pauli*. A very interesting analysis of Pauli's philosophical thinking was done by Harald Atmanspacher and Hans Primas.[10]

The core of the problem of the relationship between spirit and matter is relatively simple and can be expressed by Pauli's gemlike formula, which was so often used, in slightly different forms, in his correspondence with Fierz: *we are too deeply immersed in the seventeenth century*.[11] In other words, if Western civilization persists on the dangerous path of a too-brutal separation of the subject from the object, of science from religion, of physics from metaphysics, of the causal from the acausal, a catastrophe will be imminent.

THE MOST IMPORTANT TASK OF OUR TIME: A NEW IDEA ABOUT REALITY

Jung, Pauli, and Lupasco called into question the creation of a new model of Reality in agreement with the psychological experience on the one hand and on the other, with the microphysical experience.

The next step was almost inevitable: Jung and Pauli recognized *the existence of an isomorphism between the quantum world and the world of the psyche* in different ways. In France, slightly ahead of Jung and Pauli, Lupasco reached the same conclusion.[12] In fact, Jung, Pauli, and Lupasco called into question the entire evolution of the modern world from Galileo to the present day. But this calling into question was not negative: it implied a return to the sources of modernity, a return that was enriched with all the experience of the technoscience adventure.

The model of Reality proposed by Pauli can be summarized in Laurikainen's image[13]: the observer is separated from the real by the wall of the unconscious. The difference between Pauli and Bohr's model is very clear: the role of Bohr's measurement devices is replaced with that of Pauli's *psyche* of the observer.

At first glance, Pauli's real is virtually identical to d'Espagnat's "veiled real." But there is a major difference: the isomorphism between the mental world and the quantum world is missing in d'Espagnat. In a letter to Fierz on March 5, 1957, Pauli wrote that it was reasonable to think that even so-called "inert" matter would contain some weak psyche components.[14] Such considerations would certainly have been rejected by d'Espagnat, but they are present in Lupasco's work.[15]

NEW PERSPECTIVES IN THE TERNARY-QUATERNARY DEBATE

It is obvious that a ternary, or triad, does not necessarily imply the included middle. For example, the triad (father, mother, child) can be decomposed, in its succession in time, into three binaries: (father, mother), (mother, child), and (father, child). Similarly, the Hegelian triad (thesis, antithesis, synthesis) is decomposed in turn into two binaries: (thesis, antithesis) and (antithesis, synthesis).

The included middle necessarily has a paradoxical character to the extent that it involves the unification of two mutually exclusive contradictories (A, non-A).

Jung emphasized well this paradoxical character when he described the role of the Holy Spirit in the Trinity: "As a 'tertium,' the Holy Spirit must necessarily be extraordinary, even paradoxical. . . . He is a *function* and, as such, the third person of the deity." In his essay on the doctrine of the Trinity, Jung also writes: "A contradictory tension results between One and Another. Or, any contradictory tension tends to *evolve* and give birth to *three*. Once at *three*, the tension is canceled, because the lost unity reappears. . . . Trinity is, therefore, the unity that develops in order to become noticeable. Three is One that became noticeable, and that, without the resolution of the contraries One and Another, would have remained in a state impossible to determine."[16] But what is the nature of this unification of opposites? Is it rational or irrational? For the field of the mind, our response is provided by Jung: "the unification of opposites at a higher level is not an issue of reason, nor is it a problem of will, but a psychological process of development that is expressed through symbols."[17]

The words "higher level" used by Jung are indeed significant. Everything happens as if physical reality and psyche reality were structured on levels and these levels were in a bi-univocal correspondence.

Three generates a huge vertigo: a vertigo of thinking for the physical world and a vertigo of the whole being for the psyche world. Speaking about the third phase of human life, Jung wrote: "Passages of this kind generally have a *numinous* character—they are about conversions, illuminations, convulsions, numbness of the destiny, and religious experiences, mystical or analogical."[18]

Jung strongly stressed what is at stake in the dilemma of the Three and Four: "What we have here is *the dilemma between what is purely imaginary and reality or actualization*. Actually—especially for the philosopher who is not content to be a good speaker—it is about an outstanding issue, as important as a moral issue, to which it is closely linked."[19]

Pauli was an enthusiastic supporter of the quaternary. Pauli believed, like Jung, in the value of the quaternary—the idea of the reutilization of intuition, of the inner life, of the imagination, which opens the gates to the fabulous reservoir of energy that is the unconscious, which is perceived as being irrational, evil, darkness.[20]

Is there any opposition between the ternary, which involves the included middle, and the quaternary? The ternary can be conceived of as an instrument of transition from one level of reality to another. There are two possible meanings of this transition, because there are two immediate levels neighboring a well-determined level of reality where the contradictories A and non-A are located. Let us name T_1 the middle that unifies the contradictories at the immediately "superior" level and T_2 the middle that unifies the contradictories at the immediately "inferior" level. Thus we can distinguish the exterior quaternary $T_1AT_2(\text{non-}A)A$ from the interior quaternary $A(\text{non-}A)T_1T_2$, which has the shape of a cross. We can attribute a symbolic meaning to this cross: the horizontal axis $A(\text{non-}A)$ may be associated with being captured in the fight between the opposites; the vertical axis T_1T_2 may be associated with two possibilities of evolution and involution; finally, the center of the cross may be associated with the starting point of an infinite ascent or of an infinite descent.

There is no real contradiction between Three and Four. The quaternaries just discussed remain in the field of Three. We are here in the field of metanumbers, not in that of numbers.[21] The real *fourth* that Jung and Pauli are speaking about appears in the transition, which is not logical but existential, from one level of reality to another.

UMBERTO ECO'S LOGICAL AND EPISTEMOLOGICAL ERROR

At a well-determined level of reality (let us call it NR_0), classical logic is valid. In particular, the *principle of identity* (A = A) is true at this level. The influence of another level of reality, highlighted by scientific theory and experience, is manifested at the considered level by mutually exclusive contradictory phenomena.

Two adjacent levels are related to the logic of the included middle, meaning that the T state present at a certain level is linked to a couple of contradictories (A, non-A) at the immediate next level. At each phase, involving two neighboring levels of Reality, *the axiom of noncontradiction is respected*. The repetitive structure of the action's logic of the included middle determines the entangling of levels and the coherence of nature. There is a perpetual, iterative, and cyclical transmutation of a T state in a couple of contradictories (A, non-A), namely, a *continuous amplification of the principle of noncontradiction*.

Umberto Eco says:

> To talk about sympathy and universal similarities means that one has already rejected the principle of noncontradiction. Universal sympathy is the effect of an emanation of God in the world, but at the origin of the emanation lies an unknowable One, the location of the contradiction itself. Hermetic thinking says that the more ambiguous and versatile our language, the more it employs symbols and metaphors and the more apt it is to name the One where the coincidence of opposites is realized. But when the coincidence of opposites triumphs, the principle of identity crashes. *Everything is connected.* The result: interpretation is infinite . . . hermetic

thinking transforms the theater of the world into a linguistic phenomenon and, in parallel, it withdraws from language any power of communication.[22]

There is, in all these considerations, a serious confusion: the coincidence of the opposites does not imply the abandonment of the axiom of identity and the axiom of noncontradiction but only the abandonment of the axiom of the excluded third, which must be replaced by the axiom of the included third. Noncontradiction and identity do not crash but rather amplify.

Therefore, interpretation is not infinite but finite, precise, and rigorous. The theater of the world is not reduced to a linguistic phenomenon, which is merely one facet of an infinitely more rich Reality. The most dangerous drift today is that of one-dimensional thinking, which reduces everything to a single level of reality. Nature today is neither magic, nor mechanical, nor dead—it is *alive*, as Pauli had masterly foreseen.

CHAPTER TWELVE

FROM THE QUANTUM WORLD TO IONESCO'S ANTITHEATER AND QUANTUM AESTHETICS

> The part of yes
> that is inside no
> and the part of no
> that is inside yes
> come out of their beds sometimes
> and join each other in another bed
> that is no longer yes or no.
> In this bed flows the river
> Of the most vivid waters.
>
> —Roberto Juarroz, *Nouvelle poésie verticale*[1]

FOR A YES OR FOR A NO

It is not known whether Nathalie Sarraute knew about the logic of contradiction. But her theater play *For a Yes or for a No* is a superb example of the logic of contradiction in everyday life.

The character H.2 tells the character H.1: "People say that I always argue with my friends . . . for reasons that nobody could understand. . . . I was convinced . . . at their request . . . contumaciously . . . I knew nothing. . . . I found out that I had a criminal record that designated me as 'the one who argues for a yes or for a no.' That gave me something to think about."[2]

In fact, we do not know why we say the words *yes* and *no*. When we say *yes*, we should, little by little, embrace the entire universe, the truth in all its nudity, stripped of our prejudices, of the conditionings of birth and education, of faiths of all kinds. And when we say *no*, we put at risk the truth, the whole truth. This is an untenable, absurd position that transforms us into puppets. Therefore, we will inevitably be convinced, contumaciously, of course. We understand why the theater is a privileged site to study the intimate mechanisms of the yes and the no.

We always argue for a yes or for a no. We even kill each other for it. What is the meaning of wars between religions and civilizations if not that of taking life for a yes or for a no? It is for a yes or for a no that people are murdered in concentration camps, adulterous women are stoned, works of art are destroyed, and books are burned in the name of one ideology or another.

The life that oscillates between yes and no, always uttered with absolute conviction, can only be that of a puppet. But who pulls the strings of these puppets? Is the modern world ready for the long initiation trip that could pull us out of this infernal circle? Artists were the first ones to understand that what is at stake is what is *between* yes and no.

At the end of her play, Nathalie Sarraute makes her characters (who are interchangeable; each can take both perspectives) utter these words:

H.1: For a yes . . . ; or for a no? [Silence.]
H.2: Yes or no?
H.1: And yet, it is not the same thing. . . .
H.2: True: Yes. Or no.
H.1: Yes.
H.2: No![3]

And thus hate continues its infinite work.

IONESCO AND THE NON-ARISTOTELIAN THEATER

Ionesco read Lupasco's work carefully, and he was certainly influenced by Lupasco's philosophy. In her book *Eugène Ionesco: Mystic or Unbeliever?* Marguerite Jean-Blain stresses Lupasco's key role in Ionesco's spiritual itinerary,[4] along with Jacob Boehme and Saint John of the Cross and in the company of *The Tibetan Book of the Dead* (*Bardo-Thödol*) and of the Christian Orthodox ritual. Ionesco carefully read both *Logic and Contradiction* and *The Principle of Antagonism and the Logic of Energy*, Lupasco's fundamental work about the included third—this mysterious third between good and evil, the beautiful and the ugly, truth and falsehood.

In *Fragments of a Journal*, Ionesco wrote: "It is precisely because the Greeks had the feeling of archetypal immutability that they must necessarily have had the feeling of the nonimmutable: Lupasco explains it very well. Nothing truly exists and nothing is thought otherwise than in opposition with a contrary, which also exists and which we repress."[5] Naturally, Lupasco's name and ideas appear in the play *Victims of Duty*,[6] produced at Théâtre du Quartier Latin and directed by Jacques Mauclair six years after the publication of *Logic and Contradiction*.

The characters in this "pseudodrama" are Choubert, Madeleine, the Policeman, Nicolas d'Eu, The Lady, and Mallot with one "t." The action occurs in a "small bourgeois interior." The name "Nicolas d'Eu" is interesting: "Eu" means "I" in Romanian. Nicolas d'Eu "is tall, has a big black beard, bloated eyes because of too much sleep, disheveled hair, wrinkled clothes, and the figure of someone who has just woken up after having slept with his clothes on." Nicolas d'Eu shares his ideas about theater with the Policeman: "I have long thought about the possibility of a renewal of the theater. How can there be something new in the theater? What do you think, Mr. Superintendent?" The policeman asks him: "A non-Aristotelian theater?" "Exactly," responds Ionesco, alias Nicolas d'Eu. And he continues:

> However, it is necessary that we take into account the new logic, the revelations brought by a new psychology . . . a psychology of antagonisms. . . . While getting inspiration from another logic and from another psychology, I would introduce the contradiction into noncontradiction, the noncontradiction into what common sense believes contradictory. . . . The principle of identity and of the unity of characters will be character, in favor of motion, of a dynamic psychology. . . . We are not ourselves . . . no personality exists. Inside us there are only contradictory and noncontradictory forces . . . . In fact, it might help you to read *Logic and Contradiction*, Lupasco's excellent book. . . . Characters lose their shape in shapeless becoming. Each character is less itself than the other. . . . As for action and causality, one should not even mention them. One should completely ignore them, at least in their old form; too rough, too obvious, and too false, like everything that is obvious. . . . There is no drama or tragedy: tragedy becomes comedy, comedy is tragic, life becomes joyful . . . life becomes joyful.

The Policeman reacts in a politically correct way: "As far as I'm concerned, I remain a logical Aristotelian, faithful to myself, faithful to my job, respectful toward my bosses. . . . I do not believe in the absurd; everything is consistent, everything can be understood . . . due to the efforts of human thinking and that of science."[7]

It is interesting to note what the Policeman says: "everything is consistent, everything can be understood . . . due to the efforts of human thinking and that of science."

The director of *Victims of Duty*, Jacques Mauclair, said on May 7, 1988, the Third Day of the Molières Awards: "Mr. Ionesco, Master, my dear Eugène, you have promised to come this evening and you came. Certainly you will always amaze us. Lupasco's disciple, whose name rhymes with yours, curiously, you reconcile logic and contradiction without any apparent difficulty. Thus, you are staying away from gossips, but you do not miss any of them; you despise honors, but you accept all of them. You have a very shiny green frock coat, a bicorne, a sword, but you attend the meetings of the Academy without wearing a tie. You insult grammar, you make vocabulary a martyr, but your picture can be found inside the Petit Larousse dictionary."[8]

Critics became aware of the role of Lupasco's philosophy in the genesis and development of the so-called "theater of the absurd," terminology rejected by Ionesco in favor of "antitheater."

Emmanuel Jacquart writes about *The Lesson*:

> Without being, like Lewis Carroll, a professional logician, Ionesco is passionate about logic whenever it amuses and surprises. The simplest form of his approach is to counteract or to cancel *the principle of identity and noncontradiction*. In this way, intellectual understanding becomes mathematical reasoning that is both inductive and deductive. Latin, Spanish, and neo-Spanish rely on "identical similarities"! Finally, characters may state and refute phrases that they advance. . . . In this upside-down world, the most simple and most complex thing are equally possible. . . . When logic affords all freedoms, causality does the same. . . . In the extreme case, Ionesco imagines a fictitious logic, which structures the world in a strange manner.[9]

But the critic who emphasized the influence of Lupasco's work on Ionesco's theater in the most pertinent way is the great American theorist of art and literature Wylie Sypher, in his book *Loss of the Self*.[10] Sypher unambiguously states: "Ionesco eliminates the laws of cause and effect on which have been built both theater and science. Instead, Ionesco accepts . . . Stéphane Lupasco's logic, whose work gives us the key to Ionesco's work about the theater."[11] Sypher starts with the observation that Ionesco, like Heidegger, was fascinated by the abyss of emptiness that embraces our existence. *Ionesco purposely wants to capture the unsustainable.*

Sypher points out that science has altered the nature of literature. Any verbal language becomes a cliché in relation to the truths captured by mathematical language. "The new mathematics, and abstract painting or the new music can no longer be expressed through verbal language," says Sypher. "Mathematics has so deeply invaded all sciences and philosophy, that it is no longer possible to express reality through a verbal language, while our world is like a game played inside topology, that is able to pretend structures beyond the borders of the old logic, structures that can no longer be described by a vocabulary, with the exception of the non-verbal one."[12]

The old logic excluded feeling. Feelings, Sypher wrote, are "unique—no feeling is exactly the same as another feeling. Therefore, our feelings are discontinuous and are not subject to any logical sequence. Moreover, feelings are beyond thinking—they cannot be rationalized. In short, the old logic was a means of excluding or reducing experience—it was not a means of apprehension of experience."[13] According to Lupasco, Sypher noted, tragedy has always had the ability to capture the absurdity of life, which logic is unable to do: tragedy describes the contradictions of our human experience. "In his own way," wrote Sypher, "Lupasco is seriously taking into account what I have always said about the tragic nature of life, he considers it seriously enough to try to enrich logic with the tragic understanding of the human experience." There is, Sypher notes, a dialectic of the comic, just as there is dialectic of the tragic: ". . . comedy has its own look on human folly."

According to Sypher, "Lupasco searches for an existential logic, a logic full of 'creative contradictions,' and considers the absolute as a threat. . . . Lupasco evokes a logic of absurdity, a logic that has something in common with those *koans* from Zen Buddhism. . . . Zen seeks a direct perception of reality, without any intellectual contamination."[14]

These considerations open an unsuspected research route for exploring art and its connection to reality, thanks to the included middle.

GREGORIO MORALES: QUANTUM AESTHETICS AND QUANTUM THEATER

By the end of the twentieth century, the quantum world and its logic of contradiction had given birth to an interesting direction in theater, literature, and painting known as "quantum aesthetics." Its principal thinker is the Spanish novelist Gregorio Morales (b. 1952). In 1994, Morales founded the group "Salon of Independents," formed by sixty Spanish writers and joined by artists and scientists from other countries. Four years later, Morales published the manifesto of this movement, *Balzac's Corpse: A Quantic Vision of Literature and Art*.[15] Of course, the publication in 1991 of the book *Art and Physics* by Leonard Shlein, a member of the Salon of Independents, prepared the way for this emergence of quantum aesthetics.[16] In 1999, the Quantum

Aesthetics Group was founded. It was formed by painters, photographers, storytellers, crafters, poets, and psychologists. Quantum aesthetics became known in the United States through the collectively written book *The World of Quantum Culture*, published in 2002.[17] The first chapter of this book was written by Morales.[18] A rich Internet site (in Spanish and in English) is consecrated to the life and work of Gregorio Morales.[19]

Quantum aesthetics is surprising because of its depth. Far from any naive transfer from physics to art and far from science fiction, it takes as basic principles the extension of Heisenberg's principle of uncertainty and of Bohr's principle of complementarity, as well as Jung's analytic psychology and David Bohm's implicate order. Those viewing an art piece are really creators of their own world. In an interesting study of Xaverio's painting, Morales writes: "each observer creates his own work of art; each investigator determines the result of the experiment."[20] In the same study, Morales refers to Eddington: "material becomes something 'mental,'" and also to Jung, when Morales speaks about "individuation." He also says, "matter is intelligent, like consciousness." For Morales, "a new strange fourth dimension opens up, which projects us far from ourselves and leads us into the forbidden and irresistible kingdom of the unknown." Morales also speaks about "the depth of hidden reality" and about the *coincidentia oppositorum*: "for the quantum artist, binary thought has been substituted for 'blurry thought.' A and non-A can exist at the same time. Here is the hidden truth of nature. Alongside our universe there are other universes. And none exclude the others. All are true." Everything is interconnected, nonseparated. But, strangely enough, in spite of Morales's frequent reference to the "fruitful union of opposites," Lupasco is never quoted in regard to his included third logic.

Gregorio Morales and the other practitioners of quantum aesthetics evoke everything from the quantum world—particles, antimatter, superstrings, and black holes—but this is done just to stimulate the activity of the imaginary. They adopt a complex and multidimensional viewpoint.

Morales even wrote a book of beautiful poems, *Quantum Song*.[21] One of them is called "Particle Power," which is very interesting, because it refers to the power of the unconscious, conceived of as "god":

> Although what we cease doing
> continues in parallel universes,
> our wings
> will not carry us there.
> The unconscious decides which way to go,
> for it knows the way beforehand.
> The unconscious is our guide,

it is our binding force.
The unconscious is our god.
We are slaves to the unknown,
to shadows,
to depths,
to the infinitesimally small.

In a penetrating study of this poem, Allan Riger-Brown remarks:

Modern science, and quantum mechanics in particular, outline a picture of "reality" which, though not identical with, or reducible to, mental phenomena (as in radical idealism), is not an unproblematic "given," an "external" object independent of the process of perception. More generally, there is a growing recognition, not only among theoreticians but also among scientists engaged in eminently practical work, of the limitations of a traditional cognitive approach based on certain deceptively self-evident dichotomies, including "objective" versus "subjective," "matter" versus "mind," and "part" versus "whole." As in the case of other abstractions (including fundamental scientific concepts such as energy, mass, momentum, inertia, etc.), such dualistic distinctions are today increasingly understood to be ultimately expedient symbols, necessary simplifications or, as Gregorio puts it in one of his poems, "imaginary coordinates" rather than descriptions of an immediately knowable, bedrock level of reality. . . . What counts, aesthetically speaking, are the poet's responses, his wonderment at the creative power and complexity of the universe; the thrill that accompanies the artistic intuition of the depths of existence, the "implicate" or "enfolded order." What counts, and we are moving here into a higher level of aesthetic exploration—the ethical level—is the sense of awe which comes from the realization that life, the emergence of increasingly complex organic forms, and (hence) of humanness and consciousness, are inbuilt in the smallest particles of matter . . . not only do we intuit the essential unity of being, but we also suddenly see the universe looking at itself, "like mirrors reflected in mirrors," through the bewildered eyes of a human being![22]

The theme of *unity of being* is central in the poems of Gregorio Morales.
In this context, the parallel movement of *quantum theater* appeared. As in the case of Peter Brook, the empty, vacuum space is crucial for receiving aesthetic information, and the spectator plays an important role in the

creation of a theatrical event; as in the case of Ionesco's theater, the unity of contradictories is also very important. Simona Modreanu pointed out the influence of Ionesco and Lupasco on quantum theater.[23] Another important point for quantum theater is the existence of different levels of Reality, involving a discontinuous space-time during theatrical representation. Pablo Iglesias Simón coined the expression "quantum surrealism" in relation to quantum theater.[24]

In France, quantum theater is illustrated by the work of the famous director Claude Régy (b. 1923), who also wrote two books of theoretical considerations concerning quantum theater.[25] His representation of *Psychosis* by Sarah Kane, in 2002, was memorable. Claude Régy puts silence, as mediator of the unknown, at the center of his work. One might also mention the famous movie(s) *Smoking / No Smoking*, directed by Alain Resnais (b. 1922), which is directly inspired by the parallel world's interpretation of quantum mechanics.

Let us finally mention that at San Jose State University, an MA thesis was written on quantum theater.[26]

Quantum aesthetics and quantum theater are exemplary cases of a fruitful cooperation between science and culture in the twenty-first century.

CHAPTER THIRTEEN

THE THEATER OF PETER BROOK AS A FIELD OF STUDY OF ENERGY, MOVEMENT AND INTERRELATIONS

Throughout his life, the greatest director of our time, Peter Brook, has been obsessed by the question of the meaning of theater and has been inevitably led to an investigation of both science and Tradition.[1] If theater springs from life, then life itself must be questioned. Understanding theatrical reality entails understanding the agents of that reality, the participants in any theatrical event: the actors, director, and spectators. For a person who rejects all dogma and closed systems of thought, Tradition offers the ideal characteristic of unity in contradiction. Tradition conceives of understanding as being something originally engendered by experience, beyond all explanation and theoretical generalization. Isn't the theatrical event itself "experience," above all else?

Even on the most superficial of levels, Brook's interest in Tradition is self-evident: one thinks of his theatrical adaptation of one of the jewels of Sufi art, Attar's *Conference of the Birds*, of his film taken from Gurdjieff's book *Meetings with Remarkable Men*, and of his subsequent work on *The Mahabharata*.

One can find a precise point of contact between Tradition and theater in Tradition's quality of vital immediacy. Brook wrote: "Theatre exists in the here and now. It is what happens at that precise moment when you perform, that moment at which the world of the actors and the world of the audience meet. A society in miniature, a microcosm brought together every evening within a space. Theatre's role is to give this microcosm a burning and fleeting taste of another world and thereby interest it, transform it, integrate it."[2]

Theatrical work, traditional thought, scientific thought: such a meeting is perhaps unusual but certainly not fortuitous. By Peter Brook's own admission, what attracted him to theatrical form as well as to the study of Tradition was precisely this apparent contradiction between art and science. So it is not at all surprising that a book such as Matila Ghyka's *Le Nombre d'Or*[3] (a discussion of the relationship between numbers, proportions and feelings) should have made such a strong impression on him.[4]

Brook's theatrical research is structured around three polar elements: energy, movement, and interrelations. "We know that the world of appearance," wrote Brook, "is a crust. Under the crust is the boiling matter we see if we peer into a volcano. How can we tap this energy?"[5] Theatrical reality is determined by the movement of energy, a movement only perceivable by means of certain relationships: the interrelations of actors and that between text, actors, and audience. Movement cannot be the result of an actor's action: the actor does not "do" a movement, it moves through him or her.

The simultaneous presence of energy, movement, and certain interrelations brings the theatrical event to life. With reference to *Orghast*, Brook spoke of "the fire of the event," which is "that marvelous thing of performance in the theatre. Through it, all the things that we'd been working on suddenly fell into place."[6] This "falling into place" indicates the *sudden* discovery of a structure hidden beneath the multiplicity of forms and apparently extending in all directions. That is why Brook believes the essence of theatrical work to be in "freeing the dynamic process."[7] It is a question of "freeing" and not of "fixing" or "capturing" this process, which explains the suddenness of the event. A linear unfolding would signify a mechanistic determinism, whereas here the event is linked to a structure that is clearly not linear at all but rather one of interrelationships and interconnections.

Event is another key word frequently recurring in Brook's work. Surely it is not simply coincidence that since Einstein and Minkowski, the same word covers a central notion in modern scientific theory (see chapter 4).

Like contemporary scientists, Brook is convinced of the materiality of energy. Describing the characteristics of "rough theater," he says: "The Holy Theatre has one energy, the Rough has others. Lightheartedness and gaiety feed it, but so does the same energy that produces rebellion and opposition. This is a militant energy: it is the energy of anger, sometimes the energy of hate."[8]

But where does this energy come from? What are the laws governing the transformation of nondifferentiated energy into a specific form of energy? Is this nondifferentiated energy the fundamental substratum of all forms? To what extent can actors and audience at a theatrical performance become implicated and integrated with the formidable struggle of energies that takes place at every moment in nature?

In Peter Brook's theatrical research, the grouping text/actor/audience reflects the characteristics of a natural system: when a true theatrical "event" takes place, it is greater than the sum of its parts. The interactions between text and actors, text and audience, and actors and audience constitute the new, irreducible element. At the same time, text, actors, and audience are true subsystems, opening themselves up to each other. In this sense, one can talk of the life of a text. As Brook has said many times, a play does not have a form that is fixed forever. It evolves (or involves) because of actors and audiences. The death of a text is connected to a process of closure, to an absence of exchange. In *The Empty Space*, we read that "A doctor can tell at once between the trace of life and the useless bag of bones that life has left. But we are less practiced in observing how an idea, an attitude or a form can pass from the lively to the moribund."[9]

Almost all of the actors' "exercises" and "improvisations" in Brook's Centre seem to aim at engendering opening and exchange. Firsthand testimonies to this effect are numerous. Brook has explicitly said himself that by means of these exercises and improvisations, the actors are trying to "get to what's essential: in other words to that point at which the impulses of one conjoin with the impulses of another to resonate together."[10] Michel Rostain describes how, during the rehearsal for *Carmen*, one singer would turn his or her back on another in order to try to recreate the gesture accompanying the other person's singing *without ever having seen it*. Actors sitting in a circle attempted to "transmit" gestures or words, and in the end, the force and clarity of internal images enabled them to be made "visible." This is genuinely precise and rigorous research work.[11]

In one exercise during the preparation for *Orghast*, each actor represented a part of a single person, including, for example, "the voice of the subconscious."[12] In another, actors took part in the recitation of a monologue from a Shakespearean text, delivering it as a round for three voices: "suddenly the actor bursts a barrier and experiences how much freedom there can be within the tightest discipline."[13] And that is what it is essentially about: the discovery of freedom by submitting oneself to laws that permit an opening into the "unknown," into a relationship. Exercises and improvisations offer the possibility of "interrelating the most ordinary and the most hidden levels of experience, of discovering potentially powerful equivalences between gestures, words and sounds. In this way, words, the usual vehicle of signification, can be replaced by gestures or sounds."[14] "Going into the unknown is always frightening. Each letter is the cause of the letter that follows. Hours of work can come out of ten letters, in a search to free the word, the sound. *We are not trying to create a method, we want to make discoveries*."[15] So exercises and improvisations have little particular value in themselves, but they facilitate a tuning of the theatrical "instrument" that is the actor's being and a circulation of "living dramatic flow"[16] in the actors as a group. The theatrical "miracle" is produced *afterwards*, in the active presence of the audience, when an opening into the "unknown" can be mobilized more fully. But what is the nature of this "unknown"? Is it another name for the unity of all existing levels of Reality?

Would it be possible to discuss a theatrical "event" without immersing oneself in an experience of time? One might argue that the essence of a Peter Brook theatrical event is in its *suddenness*, in its unforeseeable nature (in the sense of the impossibility of precise reproduction at will). Brook says that "The special moments no longer happen by luck. Yet they can't be repeated. It's why spontaneous events are so terrifying and marvelous. They can only be rediscovered."[17] Meaning "never belongs to the past"[18]: it appears in the mystery of the present moment, the instant of opening to a relationship. This "meaning" is infinitely richer than that of classical realism. At fleeting moments, great actors touch upon this new kind of "meaning." In Paul Scofield, for example, "instrument and player are one—an instrument of flesh and blood that opens itself to the unknown. . . . It was as though the act of speaking a word sent through him vibrations that echoed back meanings far more complex than his rational thinking could find."[19] The "miracle" of Peter Brook's theatrical work resides in precisely this sense of the moment, in the liberation of energies circulating in harmonic flux, incorporating the spectator as active participant in the theatrical event.

Texts by Chekhov, "the dramatist of life's movement,"[20] or by Shakespeare enable every dimension of Brook's theater work to be revealed. In *The Cherry Orchard*, there are specific moments when apparently banal words and gestures fall apart, suddenly opened to another reality that one somehow feels to be the only one that counts. A flow of energy of a new quality starts to circulate, and the spectators are carried off to new heights, in a sudden confrontation with themselves.

Another remarkable meeting point between Peter Brook's theatrical work, traditional thought, and quantum theory is in the ternary structure of Brook's theatrical space, founded on the logic of contradiction.

The role of contradiction is apparent in the changes of direction Brook himself has chosen throughout his career, through Shakespeare, commercial comedy, television, cinema, and opera: "I've really spent all my working life in looking for opposites," Brook suggested in an interview with *The Times*. "This is a dialectical principle of finding a reality through opposites."[21] He emphasizes the role of contradiction as a means of awakening understanding, taking Elizabethan drama as an example: "Elizabethan drama was exposure, it was confrontation, it was contradiction and it led to analysis, involvement, and recognition and, eventually, to an awakening of understanding."[22] Brook points out the constructive role of negation in the theater of Beckett:

"Beckett does not say 'no' with satisfaction: he forges his merciless 'no' out of a longing for 'yes,' and so his despair is the negative from which the contour of its opposite can be drawn."[23]

Contradiction also plays a central role in the works of Shakespeare, passing through many levels of consciousness: "What enabled him technically to do so, the essence, in fact, of his style, is a roughness of texture and a conscious mingling of opposites."[24] Shakespeare remains the great ideal, the summit, an indelible point of reference for a possible evolution in theater: "It is through the non-reconciled opposition of Rough and Holy, through an atonal screech of absolutely unsympathetic keys that we get the disturbing and the unforgettable impressions of his plays. It is because the contradictions are so strong that they burn on us so deeply."[25] Brook sees *King Lear* as a vast, complex, coherent poem, attaining cosmic dimensions in its revelation of "the power and the emptiness of nothing—the positive and negative aspects latent in the zero."[26]

Contradiction is the *sine qua non* of successful theatrical performance. Five centuries ago, Zeami (1363–1444), one of the first great masters of the Noh (his treatise is known as "the secret tradition of the Noh"), observed: "Let it be known that in everything, it is at the critical point of harmonic balance between *yin* and *yang* that *perfection* is located. . . . If one was to interpret *yang* in a *yin* way, or *yin* in a *yang* way, there could be no harmonizing balance, and perfection would be impossible. Without perfection, how could one ever be interesting?"[27]

Zeami elaborated on a law called *johakyu*, to which Peter Brook often refers. *Jo* means "beginning" or "opening," *ha* means "middle" or "development" (as well as "to break," "crumble," "spread out"), and *kyu* means "end" or "finale" (as well as "speed," "climax," "paroxysm"). According to Zeami, it is not only theatrical performance that can be regarded in terms of *jo*, *ha*, and *kyu* but also every vocal and instrumental phrase, every movement, every step, every word.

Zeami's comments are still vitally relevant to us today. One can easily imagine, for example, the boredom provoked by the performance of a tragic play that begins in climactic paroxysm and then develops through interminable expositions of the causes of the drama. At the same time, it would be possible to undertake a detailed analysis of the unique atmosphere created in the plays staged by Peter Brook as the result of conformity with the law of *johakyu*—in the structural progression of these plays, as well as in the actors' performances. But the most personal aspect of Brook's theater work lies in his elaboration and presentation of a *new ternary structuring*.

Brook's theatrical space could be represented by a *triangle*, with the baseline for the audience's consciousness and the two other sides for the inner life of the actors and their relationships with their partners. This ternary configuration is constantly present in both Brook's practice and his writings. In everyday life, our contacts are often limited to a confrontation between our inner life and our relationships with our partners: the triangle is mutilated, for its base is absent. In the theater, actors are obliged to confront "their ultimate and absolute responsibility, the relationship with an audience, which is what, in effect, gives theatre its fundamental meaning."[28]

Harmony between body, feelings, and thought is another key aspect of the work of Peter Brook.

John Heilpern, who has described the Centre International de Recherches Théâtrales (CIRT) actors' "expedition" to Africa, recalled his astonishment when he heard Peter Brook talking about the role of cerebral activity: "He pointed to the imbalance within us where the golden calf of the intellect is worshipped at the cost of true feelings and experience. Like Jung, he believes that the intellectual—the intellect alone—protects us from true feeling, stifles and camouflages the spirit in a blind collection of facts and concepts. Yet as Brook talked to me of this I was struck forcibly by the fact that he, a supreme intellectual figure, should express himself this way."[29] As someone who had branded the twentieth-century human being "emotionally constipated,"[30] Brook sheds no tears for "deadly theatre," which he considers to be the perfect expression of the cerebral element in its attempt to appropriate real feelings and experiences: "To make matters worse, there is always a deadly spectator who for special reasons enjoys a lack of intensity and even a lack of entertainment, such as the scholar who emerges from routine performances of the classics smiling because nothing has distracted him from trying over and confirming his pet theories to himself, whilst reciting his favourite lines under his breath. In his heart he sincerely wants a theatre that is nobler-than-life, and he confuses a sort of intellectual satisfaction with the true experience for which he craves."[31]

Harmony between body, feelings, and thought facilitates the development of a new quality of perception, a direct and immediate perception that does not pass through the deforming filter of cerebral activity. Thus a new intelligence can appear: "along with emotion, there is always a role for a special intelligence that is not there at the start, but which has to be developed as a selecting instrument."[32] It is not surprising that Peter Brook and the famous neurophysiologist Antonio Damasio have a privileged friendship.[33]

A lot of the exercises elaborated by Peter Brook have as their precise aim the development of this state of unity between thought, body, and feelings by liberating the actor from an overcerebral approach. In this way, actors can be organically linked with themselves and act as a unified "whole" being rather than as a fragmented one.

The most spectacular illustration of the crucial, primary role of experience in Brook's work is perhaps in the preparation for *Conference of the*

Birds. Instead of plunging his actors into a study of Attar's poem, or committing them to an erudite analysis of Sufi texts, Brook led them on an extraordinary expedition to Africa. Confronted with the difficulties inherent in crossing the Sahara desert, obliged to improvise in front of the inhabitants of African villages, the actors inexorably moved toward a meeting with themselves: "Everything we do on this journey is an exercise . . . in heightening perception on every conceivable level. You might call the performance of a show 'the grand exercise.' But everything feeds the work, and everything surrounding it is part of a bigger test of awareness. Call it 'the super-grand exercise.' "[34] Indeed, self-confrontation after a long and arduous process of self-initiation is the very keystone of Attar's poem. This kind of experimental, organic approach to a text has an infinitely greater value than any theoretical, methodical, or systematic study. Its value becomes apparent in the stimulation of a very particular "quality": it constitutes the most tangible characteristic of Brook's work.

Brook's comments on *Orghast* are as significant and valid for *Conference of the Birds*, as indeed for all of the other performances: "The result that we are working towards is not a form, not an image, but a set of conditions in which a certain quality of performance can arise."[35] This quality is directly connected to the free circulation of energies, through precise and detailed (one could even call it "scientific") work on *perception*. Discipline is inextricably associated with spontaneity, precision with freedom.

How can discipline and spontaneity be made to coexist and interact? Where does spontaneity come from? How can one distinguish true spontaneity from a simple automatic response, associated with a set of pre-existing (if unconscious) clichés? In other words, how can one differentiate between associations—perhaps unexpected, but nonetheless mechanical—with their source in what has been seen already and the emergence of something really new?

A *general principle of uncertainty* seems to be active in any process in reality. It is also necessarily active in theatrical space, above all in the relationship between audience and play. In the "formula" for theater suggested by Brook (theater = RRA: répétition, représentation, assistance), the presence—"assistance"—of an audience plays an essential role: "The only thing that all forms of theatre have in common is the need for an audience. This is more than a truism: in the theatre the audience completes the steps of creation."[36]

The audience is part of a much greater unity, subject to the principle of uncertainty: "It is hard to understand the true function of a spectator, there and not there, ignored and yet needed. The actor's work is never for an audience, yet it always is for one."[37] The audience makes itself open to the actors in its desire to "see more clearly into itself,"[38] and the performance thus begins to act more fully on the audience. By opening itself up, the audience in turn begins to influence the actors, if the quality of their perception allows interaction. That explains why the global vision of a director can be dissolved by an audience's presence: the audience exposes the nonconformity of this vision with the structure of the theatrical event. The theatrical event is indeterminate, instantaneous, and unpredictable, even if it necessitates the reunion of a set of clearly determined conditions. The director's role consists of working at great length and in detail to prepare the actors, thus enabling the emergence of the theatrical event. All attempts to *anticipate* or *predetermine* the theatrical event are doomed to failure: the director cannot substitute him- or herself for the audience. The triangle comprising the "inner life of the actors—their relationships with their partners—the audience's consciousness" can only be engendered at the actual moment of performance. The collective entity that is the audience makes the conciliatory element indispensable in the birth of the theatrical event. An audience's "true activity can be invisible, but also indivisible."[39] However invisible it is, this active participation by the audience is nonetheless material and potent: "When the Royal Shakespeare Company's production of *King Lear* toured through Europe, the production was steadily improving. . . . The quality of attention that this audience brought expressed itself in silence and concentration: a feeling in the house that affected the actors as though a brilliant light were turned on their work."[40] So it is evident why Brook's research work tends toward "a necessary theatre, one in which there is only a practical difference between actor and audience, not a fundamental one."[41] The space in which the interaction between audience and actors takes place is infinitely more subtle than that of ideas, concepts, prejudices, or preconditioning. The quality of the attention of both audience and actors enables the event to occur as a full manifestation of spontaneity. Ideally, this interaction can transcend linguistic and cultural barriers. CIRT actors can communicate just as well with African villagers, Australian aborigines, or the inhabitants of Brooklyn: "Theatre isn't about narrative. Narrative isn't necessary. Events will make the whole."[42]

Many of the confusions concerning the problem of "spontaneity" appear to have their source in a linear, monodimensional conception of the theatrical event. One can easily believe in the existence of laws such as Zeami's *johakyu*,[43] but that is insufficient for understanding how a theatrical event can take place through the *transition* between the different elements of *johakyu*. If one limits oneself to a strictly horizontal view of the action of *johakyu* (*jo*, the beginning; *ha*, the development; *kyu*, the ending), it is impossible to understand how one might arrive, for example, at the ultimate refinement of the *ha* part of *ha* or reach a climax in the *kyu* part of *kyu*.

What can produce the dynamic "shocks" necessary for the movement not to stop, not to become blocked? How can the necessary continuity of a theatrical performance be reconciled with the discontinuity inherent in its different components? How can one harmonize the progression of the play, the actors' work, and the perception liberated in the audience?

In other words, horizontal movement is meaningless by itself. It remains on the same level forever; no information is forthcoming. This movement only acquires significance if it is combined with an *evolutionary* dynamic. It is as if each phenomenon in reality were subject at every moment to two contradictory movements that are in two opposing directions: one ascending, the other descending. As if there were two parallel rivers, flowing with considerable force in two opposing directions: in order to pass from one river to the other, an external intervention—a "shock"—is absolutely essential. This is where the full richness of the significance of the notion of "discontinuity" is revealed.

But in order for this "shock" to be effective, a certain concordance or overlap must exist between the "shock" (which in itself is subject to the law of *johakyu*) and the system upon which it is acting. Therefore it becomes clear why each element of *johakyu* must be composed of the three other elements—in other words, why there has to be a *jo-ha-kyu* sequence within the *jo*, the *ha*, and the *kyu*. These different components enable interaction between the different systems to take place.

Therefore, in order for a harmonious movement to appear, a new dimension must be present: *johakyu* is active not only horizontally but also vertically. If each element (*jo*, *ha*, and *kyu*) is composed in turn of three other elements, we obtain *nine* elements, two of which represent a sort of "interval." One of these is filled by the "shock" enabling the horizontal transition to take place, the other by the "shock" enabling the vertical transition to take place.

When one considers this two-dimensional vision of the action of *johakyu*, Peter Brook's insistence on the audience's central role in a theatrical event becomes clearer. The audience can follow the suggestions proposed to it by the script, the actors, and the director. The first interval—between *jo* and *ha*—can be traversed by means of a more or less automatic exchange, and the play can continue its horizontal movement. But the audience also has its own irreducible presence: its culture, its sensitivity, its experience of life, its quality of attention, the intensity of its perception. A "resonance" between the actors' work and the audience's inner life can occur. Therefore the theatrical event can appear fully spontaneous, by means of vertical exchange—which implies a certain degree of will and of awareness—thereby leading to something truly new, not pre-existent in theatrical performance.

The ascent of the action of *johakyu* toward the play's summit—the *kyu* of *kyu*—can therefore take place. The second interval is filled by a true shock that allows the paradoxical coexistence of continuity and discontinuity.

This analysis could be further refined by taking into account the treelike structure (it is neverending) of *johakyu*. Different levels of perception, structured hierarchically in a qualitative "ladder," could be discovered in this way. There are levels of spontaneity just as there are levels of perception. The "quality" of a theatrical performance is determined by the effective presence of these degrees.

The vertical dimension in the action of *johakyu* is associated with two possible impulses: one ascending (evolution), the other descending (entropic involution). The ascending curve corresponds to a densification of energy, reflecting the tendency toward unity in diversity and an augmentation of awareness.

But one might well conceive of a *johakyu in reverse*, such as appears, for example, in the subject of Peter Brook's film *Lord of the Flies*, in which one witnesses the progressive degradation of a paradise to a hell. An ideal, innocent space exists nowhere. Left to themselves, without the intervention of "conscience" and "awareness," the "laws of creation" lead inexorably to fragmentation and, in the final instance, to violence and destruction. In this way, spontaneity is metamorphosed into mechanical process.

It should be noted that *spontaneity* and *sincerity* are closely linked. The usual moral connotation of *sincerity* signifies its reduction to an automatic function based on a set of ideas and beliefs implanted into the collective psyche in an accidental way through the passage of time. In this sense, "sincerity" comes close to a lie, in relation to itself. By ridding ourselves of the ballast of what does not belong to us, we can eventually become "sincere": recognizing laws, seeing ourselves, opening ourselves to relationships with others. Such a process demands work, a significant degree of effort: sincerity must be learnt. In relation to our usual conception of it, this kind of sincerity resembles insincerity: "with its moral overtones, the word [sincerity] causes great confusion. In a way, the most powerful feature of the Brecht actors is the degree of their *insincerity*. It is only through detachment that an actor will see his own clichés."[44] The actor inhabits a double space of false and true sincerity, the most fruitful movement being an oscillation between the two: "The actor is called upon to be completely involved while distanced—detached without detachment. He must be sincere, he must be insincere: he must practice how to be insincere with sincerity and how to lie truthfully. This is almost impossible, but it is essential."[45]

The actor's predicament is reminiscent of Arjuna's perplexity when confronted with the advice that Krishna gives him in the *Bhagavad Gita*:

to reconcile action and nonaction—paradoxically, action undertaken with understanding becomes intertwined with inaction.

At every moment, actors are confronted with a choice between acting and not-acting, between an action visible to the audience and an invisible action linked to their inner life. Zeami drew our attention to the importance of *intervals of noninterpretation* or "*nonaction*," separating a pair of gestures, actions, or movements:

> It is a spiritual concentration that will allow you to remain on your guard, retaining all of your attention, at that moment when you stop dancing or chanting or in any other circumstances during an *interval* in the script or in the mimic art. The emotion created by this inner spiritual concentration—which manifests itself externally—is what produces *interest* and enjoyment. . . . It is in relation to the *level of nonconsciousness* and selflessness, through a mental attitude in which one's spiritual reality is hidden even from oneself, that one must forge the link between what precedes and what follows the intervals of *nonaction*. This is what constitutes the inner strength that can serve to reunite all ten thousand means of expression in the oneness of the spirit.[46]

The intervals of noninterpretation ask for an inner *silence* that is a sort of penitence. Silence plays an integral part in Brook's work, beginning with the research into the interrelationship of silence and duration with his Theater of Cruelty group in 1964 and culminating in the rhythm punctuated with silences that was indefinitely present in his last performances: "In silence there are many potentialities: chaos or order, muddle or pattern, all lie fallow—the invisible-made-visible is of sacred nature."[47] Silence is all-embracing, and it contains countless "layers."[48]

In fact, events and silence constitute the fabric of any theater performance. Silence comes at the end of action, as it does in *The Conference of the Birds*: "A beautiful symbolic opposition is drawn between the black of the mourning material and the hues of the puppets. Color disappears, all sparkle is suppressed, silence is established," observes Georges Banu.[49] The richness of silence confuses, embarrasses, and disturbs, and yet it is joy that is hidden within it, that "strange irrational joy" that Brook detected in the plays of Samuel Beckett.[50]

It is no coincidence that the words *empty space* form the title of one of the two books on theater Brook has published. One must create emptiness, silence within oneself, in order to permit the growth of reality to its full potential.

In *The Empty Space*, Brook wrote: "Most of what is called theatre anywhere in the world is a travesty of a word once full of sense. War or peace, the colossal bandwagon of culture trundles on, carrying each artist's traces to the ever mounting garbage heap. . . . We are too busy to ask the only vital question which measures the whole structure: Why theatre at all? What for? . . . Has the stage a real place in our lives? What function can it have? What could it serve?"[51]

The question is still being asked.

Peter Brook created a new form of theater—a *theater of spirituality*—which renews the ancient sacred roots of theater. Another major director (and former pupil of Peter Brook), the American-Romanian Andrei Serban, is the cocreator of this new form of theater.[52] The theater of spirituality nourishes our lives, so eager for meaning.

CHAPTER FOURTEEN

FROM CONTEMPORARY SCIENCE TO THE WORLD OF ART

ANDRÉ BRETON AND THE LOGIC OF CONTRADICTION

During an interview in 1952, André Parinaud asked André Breton: "Do you not think that the physical sciences, which have been changing the structure of our conception of the world for the past fifty years, have influenced the work of artists?" André Breton replied: "I witness that artists have been unanimous, or almost, regarding their disinterest in quantum theory and in Heisenberg's mechanics. . . . It is true that the links are not totally broken, due to some thinkers such as Gaston Bachelard, to whom surrealism owes, in particular, the fact that it can reflect itself in a 'supra-rationalism,' and Stéphane Lupasco, who reinstated emotionality in the philosophical field and, moreover, at the end of *Outline to a New Theory of Knowledge*. By fulfilling, specifically, the desire of poets, he substituted the principle of noncontradiction for the principle of contradictory complementarity."[1] Breton refers here to *Essai d'une nouvelle théorie de la connaissance*, which is the title of the second volume of Lupasco's first book, *Du devenir logique et de l'affectivité*, written in 1935.[2]

André Breton (1896–1966) and Stéphane Lupasco met and respected each other. André Breton was among the first people who realized the importance of Lupasco's philosophy for the understanding of art. In an interview published in Madrid in 1950, Breton said: "It is well-known today that surrealism aimed at nothing but convincing the mind to overcome the barrier posed by opposites such as action and dream, reason and madness, sensation and representation, etc., which represent a major obstacle in Western thinking. In its sustained effort in this regard, surrealism did not cease to evaluate the elements of support that it found in the dialectics of Heraclitus and Hegel (taking into account the recent revisions made by Stéphane Lupasco's works)."³ Breton refers here to Lupasco's book *Logique et contradiction*,⁴ for which he had a profound admiration. Lupasco's work focuses on what may be called *transfigured antinomies*. It is precisely what used to bother Jean-Paul Sartre: "For Sartre, who places reality in the very field of consciousness," wrote Mark Polizzotti, "Breton's attempts to unify the conscious action and the unconscious action 'at some point of the spirit' where dualities 'cease to be perceived as contradictions' are, at best, frivolous." "The deep source of the misunderstanding," wrote Sartre, "consists in the fact that surrealism does not care about the dictatorship of the proletariat and it sees in revolution, as pure violence, the absolute end, while communism aims to take over power and it uses this goal to justify the blood that would be spilled."⁵

Breton's appraisals are even clearer in *Anthology of Black Humor*, where, in the chapter dedicated to Jean-Pierre Duprey, he refers again to *Logic and Contradiction*. "The idea of the preeminence of light over obscurity," Breton says, "can undoubtedly pass for a relic of heavy Greek philosophy. In this regard, I give paramount importance and one of the highest liberating virtues to the objection raised by Mr. Stéphane Lupasco concerning Hegel's dialectical system, which gives more tribute to Aristotle than is deserved. . . . In addition, in the eyes of artists, this work will present the huge interest of establishing and clarifying the connection, 'one of the most enigmatic' that exists between logical values and their contradiction, on the one hand, and emotional data, on the other."⁶

Indeed, *Logic and Contradiction* is a surprising meditation on *feelings* that reveals itself as a special world in relation to the logical world. "In all knowledge, in the very structure of any cognitive process, there is a *subject of ignorance* and there is an *object of knowledge*," wrote Lupasco. "And therein lies the source of the unconscious logic and that of the conscious logic that characterize the logical experience." Feelings are directly related to the interaction between subject and object. In the absence of feelings, the subject becomes an object. Therefore, feelings have an "external nature in relation to the nature of the logical universe." This "is nothing of what the experience or the logic represent; it is neither identical, nor diverse, nor extensive, nor intensive, nor temporal . . . in this way it escapes any conceptualization." Lupasco adds: "Feelings, like being, are granted alike to the existential configurations of the logical nothingness."⁷

André Breton certainly meditated much on the third chapter, "The Logic of Art or the Aesthetic Experience." For Lupasco, art is the search for "the existential logical contradiction" and aesthetics is "the quantum science in germ, in swaddling clothes," a "consciousness of consciousness." The art of a certain era uses the cognitive material of that phase, and therefore art history is grafted on the "history of knowledge that is about logical becoming." Lupasco prophesies that art "will largely blossom toward the end of a 'utilitarian' development that is a logical development of cognition."[8] This prophecy must have had a special resonance for André Breton.

Therefore, it was natural for Breton to ask Stéphane Lupasco to answer, alongside Martin Heidegger, Maurice Blanchot, André Malraux, Georges Bataille, René Magritte, Roger Caillois and Joyce Mansour, the famous survey published in the *The Magical Art*. In a postcard dated September 16, 1955, André Breton wrote to Lupasco, "Even if the philosophers had pampered me (and that is far from the case), your opinion would still be the most valuable to me."[9]

In his text, Lupasco refers to "the cosmological conflict between subject and object" and shows that he believes that magic is bathed in mystery "because the mystery of mysteries is precisely this absolute, indisputable reality of affective states." We can penetrate this mystery, as Lupasco suggests, "in a magical way" through a "jumping causality." "The magic psyche . . . with its magical shadows, these shadows penetrated by the ontological mystery of emotionality"[10] corresponds to Lupasco's third matter.

The inscription that André Breton wrote on a copy of *The Magical Art* that he sent to Lupasco is both moving and disturbing: "For Stéphane Lupasco, the first one qualified to enter these 'magical shadows of the psyche' (although I am terrified by the thought that he could lend himself to the glorification of inquisitorial measures!)."[11] What could the "inquisitorial measures" in question possibly be?

GEORGES MATHIEU AND ARISTOTLE'S CAGE

In the inscription to Lupasco, Breton refers to the exhibition "Commemorating Ceremonies of the Second Condemnation of Siger de Brabant," organized by Mathieu and Hantaï in the five rooms of the Kléber Gallery in Paris March 7–27, 1957. The program of these "ceremonies"[12] holds the memory of a stunning suite of scenerios, staging, works of art, documents, musical programs, conferences, statements, and ceremonies that were held in a rapid deluge for three weeks. Several important personalities—Carl Gustav Jung, T. S. Eliot, Karl Jaspers, Gabriel Marcel, Jean Paulhan, and Stéphane Lupasco—participated in this event, which had a great deal of media coverage. The purpose of this unusual, yet very carefully prepared event was to "call into question the foundations of our Western civilization, since the invasion in Europe and therefore in France, of Aristotle's thinking," as Siger de Brabant interpreted it. "These events, which display Western cultural and spiritual development from the Edict of Milan in 313 to 1944, reactivate the convictions capable of strengthening Georges Mathieu's irreversible commitment to a 'renaissance.'"[13]

Stéphane Lupasco was present in that part of the program entitled "The popular cycle 1832–1944 (From the *Mirari vos* encyclical to the revolution in 1944–1946)," particularly in the section "key moments remembered." In one of the halls, there was a huge portrait of Lupasco. At the foot of this portrait, on the left, Mathieu placed a small bust of Aristotle in a cage enclosed with a grill. All around, on the walls, on the ceiling, and on the floor, there were many photographs, objects, and inscriptions. Mathieu was right when he perceived the Aristotle/Lupasco opposition. This was a legacy of the Heraclitus/Parmenides opposition, which crosses the entire history of Western philosophy.[14]

Was this challenging event capable of justifying Lupasco's excommunication? In any case, it established a silence in the relationship between Breton and Lupasco.

Georges Mathieu (b. 1921) met Lupasco in 1953 as a result of his passionate reading of Lupasco's book *The Principle of Antagonism and the Logic of Energy*.[15] Mathieu is the one who opened the doors to the world of art for Lupasco—artists, journalists, philosophers, and writers: Dali, Michaux, Revel, Axelos, Abellio, Alvard, Parinaud, and Bourgois. Mathieu asked Lupasco to contribute to the lavish and widely distributed magazine *United States Lines: Paris Review* and to the magazine *Ring des Arts*.[16]

Mathieu's theoretical writings are largely imbued with Lupasco's ideas. Thus, his theory about lyrical abstraction, the continuation of the very exciting "embryology of the signs," is strongly influenced by Lupasco.[17] Indeed, Mathieu does not hide this influence: his writings are crowded with references to Lupasco. Mathieu proceeds thus not only because of his admiration for Lupasco's work but also by virtue of a deep kinship with his own creative acts. Mathieu's work as a painter is a true incarnation of contradictory logic. The passion and rage of the creative act seem to emerge from the included third, through an increased acknowledgment of feelings, this hidden third that haunted Mathieu during his entire life.

Lupasco wrote about Mathieu's work. In a hard to find text now available in the superb volume that appeared in 2003, *Mathieu: 50 ans de creation*, Lupasco wrote: "If the Greek Sphinx—Greek par excellence—immediately devoured passersby who were not able to decipher his enigma, Mathieu kills them immediately, be it only for trying to decipher it."[18]

SALVADOR DALI AND NUCLEAR MYSTICISM

In one of Alain Bosquet's interviews with Salvador Dali (1904–1989), we witness this surprising dialogue:

A. B.: With what famous man would you like to spend the night?

S. D.: Unfortunately, he's dead. Max Planck.

A. B.: That's an odd choice.

S. D.: He discovered the black body in physics.

A. B.: The *quantum* man?

S. D.: That's right.[19]

In his *Anti-Matter Manifesto*, Dali evokes another founding father of quantum mechanics—Werner Heisenberg: "In the surrealist period, I wanted to create the iconography of the interior world—the world of the marvelous, of my father Freud. I succeeded in doing it. Today the exterior world, that of physics, has transcended that of psychology. My father today is Dr. Heisenberg."[20] In the same year, 1958, Dali gave a talk on Heisenberg at Théâtre de l'Etoile in Paris.

Dali's interest in Einstein's theory of relativity is well-known. The fact that space and time are intertwined might have intrigued Dali. In her masterly book about the fourth dimension in modern art, Linda Dalrymple Henderson clearly showed the influence of the theory of relativity in Dali's paintings (especially in the series of "soft watches").[21] But the influence of the quantum world is virtually ignored in critical literature about Dali.

Numerous paintings and Dali's theological writings (such as, for example, that devoted to the dogma of the bodily assumption of the Virgin Mary into heaven, promulgated on November 1, 1950) clearly demonstrate that Dali was familiar with quantum mechanics.[22]

Dali was in close relationship with Lupasco, who certainly influenced the way in which Dali integrated the quantum vision into his work.

Lupasco dedicated a study, which appeared in a special issue ("Homage to Salvador Dali") of the magazine *XXe siècle*,[23] to Dali. Lupasco notes that traditionally, human beings have considered that divine forces come from the heights, from a world above us, a "supraworld." Lupasco even calls this faith "surrealism." But for Lupasco, Dali's case showed the full importance of the quantum "infraworld." Dali was a visionary who captured the quantum movement, the continuous creation and annihilation of shapes, confronting unusual entities with no counterparts in our own world that are nevertheless real in their own world. This is why Lupasco suggests the term *subrealism* for the description of Dali's artistic creation.

Dali made a very important symbolic gesture when, in 1978, he invited Lupasco, as his favorite philosopher, to the television show *The Thousand and One Visions of Dali*.[24] During this show, Dali told Lupasco: "You have just masterfully darkened the light!"[25]

The period in Dali's work after World War II that coincided with the return of Dali to the Catholic Church is known as "Nuclear Mysticism." Dali had the ambition of reinterpreting Christian religion in terms of quantum mechanics, a highly heretical view, of course. In his "Mystical Manifesto," Dali clearly formulates his aim of a building a metaphysical spirituality founded on quantum physics.[26]

The properties of the quantum world particularly stimulated Dali's imagination. He was convinced that quantum particles were angelic elements.[27] In "Galatea of the Spheres," a portrait of Dali's wife Gala, we see the Madonna-like face dematerializing into atomic particles represented as spinning spheres in a luminous sky. This kind of quantum vision can be found in many other paintings, such as "Raphaelesque Head Exploding" (1951), "Madonna of Port-Lligat" (1949), "Corpuscular Madonna" (1952), "Lapis Lazuline Corpuscular Assumpta" (1952), "Corpus Hypercubus" (1954), "Leda Atomica" (1947–1949) and "Saint Surrounded by Three Pi-Mesons" (1956). The basic ingredients in these paintings are *quantum discontinuity* and the omnipresence of the *quantum vacuum* (see chapters 4–8), which deeply impressed Dali.

One may question how much Dali knew about science and, in particular, quantum physics. This question is, in fact, irrelevant: Dali's work was not science but painting. His imagination was stimulated by quantum physics and engendered fabulous paintings. In other words, Dali did not illustrate quantum physics or catastrophe theory but discovered isomorphism with art in his subconscious mind. He was very fascinated with the idea of a possible *unity of knowledge*: "artists should have some notions of science in order to tread a different terrain, which is that of unity."[28] One thing is certain: his library contained hundreds of books in different scientific fields, with Dali's notes in the margins. At the end of his life, he was interested in Stephen Hawking's book *A History of Time*, and when he died in 1989, this book was found near his bed, together with books by Matila Ghyka and Erwin Schrödinger.[29]

His personal relations with Matila Ghyka and René Thom are also intriguing.

Matila Ghyka (1881–1965) became internationally famous in 1931, when he published the influential book *Le nombre d'or* (The Golden Number).[30] Dali's personal library included two books by Matila Ghyka, *The Geometry of Art and Life*[31] and *Esthétique des proportions dans la nature et dans les arts*,[32] both heavily annotated by Dali.

Dali met Ghyka in 1947, in Los Angeles, where Ghyka was teaching esthetics at the University of Southern California, and a fruitful collaboration between them began. At that moment, Dali was working on "Leda Atomica" and writing *50 Secrets of Magic Craftsmanship*.[33] It was Ghyka who computed, for "Leda Atomica," the proportions of the pentagon in which the woman represented in the painting (Gala, Dali's wife) would be inscribed. "Madonna of Port-Lligat," "Corpus Hypercubus," and the astonishing "Sacrament of the Last Supper" (1955) were also realized based on aesthetic considerations elaborated by Ghyka. Concerning the last two paintings, it is interesting to note that Dali drew a first sketch of these paintings on his copy of *Esthétique des proportions dans la nature et dans les arts*. In his book, Ghyka supplied a four-dimensional projection of a cube, a hypercube. Also, as David Lomas points out in his penetrating study " 'Painting is dead—long live painting': Notes on Dali and Leonardo":

> His makeover of the *Last Supper* is a heavy-handed object lesson in *divina proportione*. The cassocks of the priests, their heads bowed in prayer, form a repeated series of pentagons, and the overall dimensions of the canvas are those of a golden section rectangle. Dominating the upper part of the work, in a daring departure from the original, is a giant three-dimensional version of one of Leonardo's drawings for Pacioli, the skeletal armature of which opens window-like onto a vista formed by the sea and sky of Port Lligat. The geometric figure looming over the scene like a strange hallucination is a dodecahedron of the *vacuous* type, about which Dali had quizzed Ghyka in an exchange of letters in 1947.[34]

Ghyka helped Dali to better understand Leonardo's technique of painting. Salvador Dali's book *50 Secrets of Magic Craftsmanship*, written in 1948, is full of references to Matila Ghyka, and it is very instructive for understanding the deep seriousness of Dali in treating mathematical and geometrical matters.[35]

Dali's relationship with the famous French mathematician René Thom (1923–2002) was also very interesting. René Thom received the Fields medal in 1958, and he is widely known as the founder of "catastrophe theory," elaborated in 1968. This theory fascinated Dali and inspired his last series of paintings. In 1979, when he was received as a member of the Institut de France (Académie des Beaux-Arts),[36] in his acceptance speech, Dali mentioned Thom's theory as being the most beautiful aesthetic theory.[37] In 1983, Dali painted the "Enlevement topologique d'Europe: Hommage à René Thom." (Topological Abduction of Europe: Homage to René Thom).[38] In the bottom left corner, one can see an expression used by René Thom, "Queue d'aronde," and a mathematical formula. This painting perplexed art critics. Dali's last painting is "The Swallow's Tail: Series on Catastrophes" (May 1983).[39]

Four years before his death, a symposium, "Culture and Science: Determinism and Freedom," was organized at the Teatre-Museu Dali in Figueres by the Faculty of Physics of the University of Barcelona. Too weak to attend,

Dali watched on a television in his bedroom. He later invited some of the key speakers, including René Thom and Ilya Prigogine, to meet him personally in order to engage in further discussions.[40]

Separated in time by half a millennium, Leonardo da Vinci and Salvador Dali shared the same belief: that art and science are two faces of the same Reality.

FRÉDÉRIC BENRATH, KAREL APPEL, AND RENÉ HUYGHE

Lupasco's works primarily addressed the scientific world. Lupasco was greatly deceived by scientists regarding their reaction to his work, as they kept a strange silence. In fact, the explanation is quite simple: scientists abhor the word "contradiction," because the construction of science is founded on the principle of noncontradiction. Dali knew this very well when he said: "Authority will not be able to help being officially recognized after the trepanning of the small principle of noncontradiction."[41] Artists who were practitioners of contradiction received Lupasco's works with enthusiasm.

Besides the study on Dali, Lupasco also wrote about Frédéric Benrath (1930–2007)[42] and Karel Appel (1921–2006).[43] For example, he wrote about Benrath's paintings: "Antagonism is pulsating inside them. . . . ontology seems to be solely the responsibility of artists who do not look for it, who do not discuss it, but who create it; it will always be overlooked by philosophy and, further, certainly by didactic celebrations! Is it possible that such works of art may be the unusual trails, the lightning raids of another universe whose foundation they are? Benrath is among those who have the privilege to sit in its proximity."[44] When Lupasco wrote these lines, Benrath was twenty-four years old.

In Appel's painting, Lupasco sees experimental evidence of the biological matter postulated in *The Three Matters* and contemplates in it the dynamisms of the tumult and of the arborescence of the carnage. Lupasco wrote: "Appel is a painter of the infra-vital, of the infra-carnal, and his vivid touch lines are like the enemy, the very principle of his existence, claimed by the dark conception of life."[45] Appel left us a beautiful portrait of Lupasco, which at one point was available on the Internet.[46] Also, Appel's painting "Portrait of Michel Tapié and Stéphane Lupasco" can be found in the Staedelisch Museum in Amsterdam.

Lupasco's ideas have intrigued art critics. Since 1955, the famous French art critic René Huyghe has welcomed the unknown worlds that open to the spirit thanks to the understanding of art works through the logic of the contradictory.[47] He was always interested in the relationship between science and art. "Art and science," says René Huyghe in response to a question from Simon Monneret, "are not two separate worlds, with borders and custom-house officers, but two compensatory views, which complete each other." Of course, Huyghe quotes Lupasco quite often in his books, but, strictly speaking, this is not about an influence, but about a convergence: "I encountered his thinking when my mind was already made up, but I was immediately struck by the convergence of our thoughts. We started from the same findings. I admire him a lot for the conclusions he reached."[48]

Lupasco had fruitful exchanges with many other artists and art critics: Roberto Matta, Antoni Tàpies, Pierre Soulages, Henri Michaux, Pro Diaz, Nicolas Schöffer, Alain Jouffroy, Michel Ragon, and Julien Alvard.

Lupasco was convinced that his work had a scientific nature. Like a theoretical physicist, he searched for the experimental evidence of his theories. Once the three matters were inferred, Lupasco inevitably reached the idea of the existence of three worlds. Our universe is certainly a predominantly macrophysical universe. But Lupasco used to *see*, in his mind, the other two universes that for ordinary mortals seem rather like science fiction: a predominantly biological universe and a predominantly psychological universe. In the biological universe, "macrophysical systems would be in the minority, fragile and ephemeral, always contested and potentialized by macromolecular groupings, by biological configurations and structures, whose size, flexibility, range of diversity, and strength are hard to imagine." In the psychological universe, "all the consciousness pairs arise: perceptual, conceptual, and notional; the contradictory consciences will clarify each other. . . . And I will be, at the same time, a subjective ambivalent activity, a whole suite and a range of antagonistic observatory subjects, leaving the shadow of the unconscious in order to pulsate at the basis of a fine subconsciousness in a state of balance, interfering every moment with the consciousness of the consciousness while stimulating it, changing it."[49] The fascination exerted by the art world on Lupasco can be explained by the fact that the paintings of Appel and Benrath illustrate the actual existence of the biological universe and those of Dali and Mathieu illustrate the actual existence of the mental universe. They make visible the hidden universes of the logic of contradiction.

A sign of the era, the interaction between the theory of relativity, quantum mechanics, and art (especially surrealism) begins to attract the attention of art historians. In a recent book, *Surrealism, Art and Modern Science*, Gavin Parkinson offers a splendid review of this interaction.[50]

CHAPTER FIFTEEN

VISION OF REALITY AND REALITY OF VISION

An obstinate stereotype requires that scientific creation should be associated with an unwavering logical démarche, the psychological factor being present, at the most, on the accidental plane. It is true that a partially scientific, technical result is generally obtained through the rigorous development of a certain formalism. However, in the big game of scientific invention, the burning fire of the imaginary often plays a prominent role in relation to the unperturbed calculation of scientific logic. It is precisely the imaginary that will give us the hidden link between art and science. The same rules govern the imaginary in both cases.

The etymological fact that *imago* has the same root as the word *imitor* cannot hide the capacity of the creativity of the imaginary, this fabulous faculty of the possible. And where else could the dynamism of this faculty of the possible be unveiled than in science, revealing the complex relationship between image and imagination? The imaginary certainly plays an important role in the dynamic process of the concordance between human thought and the intelligence hidden in natural laws.

This field is not vastly explored. The reasons are multiple. Testimonials of the great scientific inventors on the role of the imaginary, which could represent very meaningful documents, are unfortunately very rare. There is definitely a certain embarrassment about such testimony, assessed as extra-scientific, and a certain fear of altering scientific beauty by taking into account psychological elements judged as impure and even unnecessary. Finally, once the scientific edifice is built, we can get rid of the imaginary, which nevertheless is responsible for the nature of this edifice.

One might try to go backwards, reconstructing the historical course of scientific ideas and thus obtaining clarification of the role of the imaginary in their development. But this road is too full of evident pitfalls; it is devious and, ultimately, ineffective. How can we understand the unique moment of the birth of a theory by studying a body that is already formed, fully developed, and crystallized?

Finally, an important difficulty is related to the very nature of the field under study. Here, the imaginary operates in an abstract, mathematical framework whose reasoning and complexity preclude rapid understanding. The philosopher or the psychologist who does not possess the assimilation of scientific knowledge at a high level and who has not personally experienced moments of scientific creation is disarmed by force of circumstances, almost blind in front of the beauty of the scenes displayed.

On the contrary, two papers written by scientists directly and unambiguously clarify the role of the imaginary in scientific creation. They concern the famous speech delivered by the great mathematician Henri Poincaré to the Psychological Society in Paris[1] in 1908 and the book *An Essay on the Psychology of Invention in the Mathematical Field*, written by another great French mathematician, Jacques Hadamard.[2] These two works represent the cornerstone of a direction that might prove fruitful for science, art, and philosophy.

POINCARÉ AND SUDDEN ENLIGHTENMENT

The great mathematician Henri Poincaré (1854–1912) made essential contributions to many areas of mathematics, celestial mechanics, mathematical physics, and the philosophy of science. He was the founder of algebraic topology (which he called *situ analysis*). He discovered the equations of restrained relativity (without studying their physical consequences) a few months before Einstein.

His most famous discovery was that of the Fuchsian groups. After months and months of setbacks, he abandoned his research topic. Then the solution suddenly appeared to him in the most unexpected way:

> Just at this time, I left Caen, where I was living, to go on a geological excursion under the auspices of the School of Mines. The incidents of travel made me forget my mathematical work. Having reached Coutances, we entered an omnibus to go to some place or other. The moment I put my foot on the step, the idea came to me, without anything in my earlier thought seeming to have paved its way: the transformations I had used to define the Fuchsian functions were identical with those of non-Euclidean geometry. I did not verify the idea; I would not have had time, because as soon as I took my seat in the omnibus, I went on with a conversation which had already begun, but I felt a perfect certainty. On my return

to Caen, for conscience's sake, I thoroughly checked the result.[3]

As Poincaré had not been thinking of his mathematical work during his travels, the sudden revelation of the solution and his total certainty of the validity of this solution deserve to be underlined. In fact, this experience was not unique in Poincaré's life. It was quite often repeated: "Then I turned my attention to the study of some arithmetical questions, without much apparent success and without a suspicion of any connection with my previous researches. Disgusted with my failure, I went to spend a few days at the seaside and thought of something else. One morning, walking along the bluff, the idea came to me, with just the same characteristics of brevity, suddenness, and immediate certainty, that the arithmetic transformations of ternary quadratic forms were identical with those of non-Euclidean geometry."[4] Therefore, there is a kind of selective communication, resulting from a *choice* made from an infinite number of possibilities. *But who makes the choice? There is a discernment that can see in that very place, where the watching consciousness is blind.* Paradoxically, the logical conscious effort blocks the course of this discernment, and when it stops, the requested information instantly pops up. And yet, the logical effort paves the way for the information to pop up. At the same time, one has to postulate the existence of an *unconscious logic*, a very nonlinear, nonassociative logic, able to contain the overall complexity of the problem studied. Poincaré himself speaks about *unconscious work:* "What will be striking at first is this appearance of sudden illumination, a manifest sign of a long, conscious, prior work. . . . The role of this unconscious work in mathematical invention appears to me incontestable."

Poincaré's interpretation of his own experiences is certainly not the only possible interpretation. The sudden light and the immediate certainty could be intrinsic properties of a discernment that in the normal state, that of ordinary perception, cannot operate, being absent. The previous long and conscious work gives the opportunity for this discernment to see what shows up. Thus, ordinary, associative, and timely logic helps with the manifestation of nonassociative and globalizing logic. The imaginary and the real complement each other *in a fruitful contradictory relationship*, revealing a deeper reality than that available to the sense organs.

The existence of this kind of sight is described in a precise experimental manner by Poincaré: "One evening, contrary to my custom, I drank black coffee and could not sleep. Ideas rose in crowds; I felt them collide until pairs interlocked, so to speak, making a stable combination." In this way, a subtle reversal was produced: *what was active became passive and what was passive became active.* Poincaré assists passively, as if he saw images on a screen, contemplating the avalanche of ideas and their combination, which creates truth, verified later through hard work.

Poincaré says that the subliminal self "is capable of discernment, it has tact, delicacy, it knows how to choose, to guess. . . . In a word, is not the subliminal self superior to the conscious self?"[5]

HADAMARD AND THINKING WITHOUT WORDS

The high intellectual level and the rigor of the description given by a scientist of Henri Poincaré's stature, who talked about his experiences, encouraged other scientists to disclose their testimonies. The word *rigor* in this context primarily means the absence of any philosophical or religious biases, which would do nothing but obscure and alter the experimental facts.

Thus, in 1945, another great French mathematician, Jacques Hadamard, published a very dense little book that is essential for understanding the role of the imaginary in scientific creation. One of the founders of the modern theory of equations with partial derivatives and also known for his important work on analytical functions, Jacques Hadamard showed a constant concern for physical problems and, like Poincaré, wrote numerous articles on various subjects.

Hadamard's *An Essay on the Psychology of Invention in the Mathematical Field* is largely based on the analysis of Poincaré's confession; it is a first attempt to build a theory about the role of the imaginary in science. But the major point of interest in his essay consists in the description of the experiences lived by Hadamard himself.

The author highlights *the essential role of the very short intermediate period between sleep and waking up*: "One phenomenon is certain, and I can vouch for its absolute certainty: the sudden and immediate appearance of a solution at the very moment of sudden awakening. On being abruptly awakened by an external noise, a long searched for solution appeared to me at once, without the slightest instant of reflection on my part—the fact was remarkable enough to have struck me unforgettably—and in a quite different direction from any of those that I had previously tried to follow."[6]

One notices again the *sudden*, immediate character (but always after a long period of preparation) of the actual manifestation of the revelation, without the slightest participation of ordinary logical thinking. Poincaré speaks about a sudden illumination, Hadamard about the sudden appearance of a solution, Gauss about an unsuspected light. Other confessions—those of Helmholtz, Langevin, and Ostwald, cited by Hadamard—go in the same direction.

Another important observation made by Hadamard concerns *the absence of words and even mathematical symbols* during the privileged moments of genuine scientific creation: "I insist that words are totally absent from my mind when I really think. . . . I behave in this way not only about words but even about algebraic signs. I use them when dealing with easy calculations, but whenever the matter looks more difficult, they become baggage that is too heavy for me. I use concrete representations of a quite different nature."[7]

Words and mathematical symbols are therefore embarrassing when Hadamard truly thinks. They are replaced by other representations of a different nature. But what is this other nature?

At this point another key observation of Hadamard lights the path: the vagueness of the new thinking. In this regard, Hadamard provides a masterly illustration of the theorem "the sequence of prime numbers is unlimited": "I see a confused mass; . . . I imagine a point quite distant from this confused mass. I see a second point a little beyond the first one. I see a point somewhere between the confused mass and the first point."[8] We must agree that this description is very far from any cliché about the nature of mathematical demonstrations.

The vagueness of the information received should not be confused with a lack of precision. A detail of a demonstration, a link of reasoning can undoubtedly be vague and imprecise, but the *globalism* of a problem is perceived with an unequivocal precision. *It is as if the perception of reality necessarily required abandoning the accuracy of common logic:* "Indeed, any mathematical research," says Hadamard, "compels me to build such a scheme, which is always and must be of a vague character so as not to be deceptive."[9] This will produce a new, subtle inversion: the usual logical precision confuses the researcher engaged on the path of the globalist vision. Logical precision becomes necessary *afterwards*, in the rigorous and thorough development of a demonstration stage during which the diffuse and vague could be really confusing.

Hadamard's findings on the relationships between what is vague and what is global and between what is detailed and what is precise correspond perfectly with the testimonies of painters, composers, and writers (much more numerous than those of scientists), which reveal the role of globalizing, nonlinear, wordless thinking in the genesis of an artistic work. *The similarity of nature between the functioning of the imaginary in the artistic and the scientific creation* is a remarkable fact that deserves to be examined and could eliminate much confusion.

As a rigorous scientist, Jacques Hadamard did not want to limit himself to his own experiences. He surveyed today's scientists. Einstein's response is illustrative for the reality of this thinking without words: "The words or the language, as they are written or spoken, do not seem to play any role in my mechanism of thought. The psyche entities that seem to serve as elements in thought are certain signs and more or less clear images that can be 'voluntarily' reproduced and combined. . . . These elements are, in my case, of visual and some are of muscular type. Conventional words or other signs must be sought for laboriously only in a secondary stage, when the mentioned associative play is sufficiently established and can be reproduced at will."[10]

Einstein's remark on the role of the *motor elements* in the operation of imagery is interesting, because it shows that the signal received is not necessarily visual but can be kinetic or auditory, as confirmed by other responses to Hadamard's investigation. Therefore, Gilbert Durand's hypothesis of a "close simultaneity between bodily gestures, nerve centers, and symbolic representations"[11] seems to be entirely justified.

Scientific invention is stimulated by a particular physical relaxation—many mathematicians speak of solutions found during sleep or even during the wakened state. This is also driven *by a certain attitude of letting go* of ordinary logical thinking, which favors the emergence of a different kind of attention. Conversely, the many failures of researchers who do not perceive the direct consequences of their works can be explained by the fact that they do not think *beside* as much as they need to.

Total *muscle relaxation* seems to have played an important role in the famous case of the Hindu mathematician Srinivasa Ramanujan (1887–1920).[12] With a modest education and only one available mathematics book, Ramanujan rediscovered, all by himself, the results found by European mathematicians during half a century and many important discoveries regarding numbers theory. An office clerk in Madras, he was eventually called to teach at Cambridge University. Like Hadamard, Ramanujan used to record mathematical results upon awakening. He said that the goddess Namagiri inspired the mathematical formulas in his dreams. His testimony is, unfortunately, fragmentary, and his biographers do not give much attention to the documents that might clarify one of the most spectacular manifestations of mathematical genius.

KEPLER AND THE LIVING EARTH

There is no other scientist whose imaginary has manifested with more force and fecundity than Johannes Kepler. The testimony is here direct, firsthand: in all his books, Kepler describes to the reader all the twists and turns of his thought and his demonstrations.

Kepler was interested in the rational order of the world to the point of obsession. He conceived of mathematics as the *common language* of humanity and as divinity manifested in nature. It may seem just as amazing that the vast cosmological reveries of Kepler proved to have an undoubted technical fecundity. The major aspect of Kepler's work is often obscured by most commentators because of the complexity of the subject and because of the presence of very different structures of thought from those prevailing in our time.

The lead wire of Kepler's entire work is the *archetypal image of the sphere*. In the center of the sphere, Kepler saw the symbol of the principle that is the generator of any new reality. At the surface of the sphere—and again in the interval between the center and the surface—he saw the dynamic, energetic principle that makes the manifestation possible. Beyond any mystical approach, Kepler logically states this ternary principle, which was to be found three centuries later in the philosophy of Peirce (see chapter 9).

One may wonder how an image that looks primary and poor could direct Kepler's genius for detailed calculations, for confrontations with experimental data, and especially for the theoretical realization of the unity between terrestrial physics and celestial physics. For Kepler, the symbolic ternary structure of the sphere is a principle of the dynamic unity of the world and of its invariance. For him, the sun, the fixed stars, and cosmic space are a visual materialization of this archetype. Therefore, long before Newton, Kepler showed that the tides, the mutual attraction between the Earth and the moon, and even the fall of bodies in general are phenomena of the same nature.

Kepler has a consistent respect for *corporality*, for quantity. In a world where everything is connected to everything, as a manifestation of intelligence and life, the body of Terra can only be alive: "just as the urinary bladder, the mountains, secret rivers . . . and just as in the veins of animate beings is produced the generation of blood and, together with it, sweat, rejected outside the body, in the same way, in the veins of the Earth is produced the generation of metals, fossils, and vapors of rain."[13] The food for the Earth is sea water. Terra perceives the harmonic, geometric configurations of the planets, of the sun and the moon, and enjoys this celestial harmony. It has a sense of touch, a sense of hearing, and a memory. It breathes and digests and may suffer from diseases. Finally, the Earth, just as the sun and the entire world, has a soul. In our time, Kepler would have probably seen in the vertiginous development of computer science the manifestation of the Earth's brain and perhaps would have felt close to the environmentalists, because for him, the vegetal, animal, and human coating of the planet would have been a part of the Earth's body.

A person of our time, with his or her own thinking norms, cannot see in all these imagistic visions anything but unhealthy phantasms, an awful psychosis, or a more or less poetical delusion, without any scientific justification. But how does it happen that these phantasms led Kepler toward precise calculations and toward his original approach to gravity?

The same question can be raised when taking into account the patient research conducted throughout his entire life on the possible connection between the planetary orbits or the five regular polyhedra and the structure of the musical range. "But a quite amazing fact," Kepler wrote in *The Cosmographical Mystery*, "is that I had walked directly to the target in terms of their order and that I could therefore not change it, despite the most advanced research. I could never express in words the joy that my discovery caused."[14]

However, when Kepler understood that description by regular polyhedra was leading to false experimental results, he did not hesitate to abandon it. He preserved the idea of musical harmony, and in the end, it was exactly this image of a *music of the spheres* that led him in 1619, in *The Harmony of the World*,[15] to the discovery of his famous third law regarding the movement of planets: the ratio of the square of time necessary for the planet to make a full round and the half of the major axis of the corresponding ellipse is invariant and constant. It is often said that this law was found by Kepler in an empirical way, by successive probing. His own text shows that this statement is false.

In our opinion, what underpins Kepler's whole thinking is the research on the invariance of physical laws and the creation of a scientific theory based on principles of formal, mathematical order, collation with the experimental data being performed retrospectively. In this sense, Kepler's thinking was surprisingly modern. The archetype of the sphere and all its consequences simply served as a guide on this path.

BOHR AND COMPLEMENTARITY

The significance of the idea of *complementarity* introduced by Niels Bohr in 1927 is well-known: a quantum particle can be described *approximately* either in terms of a classical corpuscle or in terms of a classic wave, but the quantum particle is *neither* particle *nor* wave. The corpuscle and the wave appear as two complementary aspects of the quantum particle that is, in this sense, *both* particle and wave.

In opposition to the meaning of the word in everyday language, this complementarity refers to the *mutually exclusive* issues represented by quantum phenomena. Of course, the deterministic causality of classical physics is not compatible with the idea of complementarity. Yet, physicists communicate the results of their experience in terms of classical logic.

Bohr's principle of complementarity was an unprecedented challenge, launched by scientific experience, to the very type of thinking that characterizes our daily life. It is therefore not surprising that the most brilliant minds of our time were hesitant to accept it. Einstein himself wrote: "It must seem a mistake to permit theoretical description to be directly dependent upon acts of empirical assertions, as seems to me to be intended, for example, in Bohr's principle of complementarity, the sharp formulation of which I have been unable to achieve despite expending much effort on it."[16] The hidden hypothesis in Einstein's statement (common as well to most people) is that of the uniqueness of logic: logic would be within the human spirit, given organically, biologically, once and forever. But quantum paradoxes seemed to indicate that logic is not unique: it has an empirical basis, it must evolve, it must change, even brutally, in order to agree with the reality of facts. The imaginary, by its freedom from memory, can make clear to us the nature of what is temporarily unthinkable. It is therefore interesting to try to uncover the way imagery was able to lead Bohr to the formulation of the complementarity principle.

Historians of science have demonstrated the considerable influence of the Western cultural atmosphere on Bohr's thinking and its primacy on philosophers such as James, Kierkegaard, and Höffding. In particular, Gerald Holton underlined the crucial role played in the formulation of the complementarity principle by Bohr's reading of William James's book *The Principles of Psychology*.[17] In an interview with Thomas Kuhn in 1962, Bohr himself explicitly acknowledged that the work of William James (and especially the chapter on schools of thought), read in 1910, was for him a true revelation.[18] According to Meyer-Abich, Jammer, and Holton, even the term *complementarity* was borrowed from James.

But apparently, William James's book has nothing to do with quantum physics. James describes cases of hysterical anesthesia, studied, for example, by Pierre Janet. Patients behave as if they had two consciousnesses that completely ignore each other: one consciousness endowed with speech, which would comply with hypnotic suggestions, triggering full anesthesia, and another more hidden, more *organic* consciousness, allowing patients to show a normal sensitivity in the areas affected by complete anesthesia. Many other phenomena have established the reality of these two consciousnesses. As James concludes: "We must agree that in some people at least, the global consciousness capable of existing can possibly split into parts which coexist, while ignoring each other, and which parcel out their knowledge objects. To put in agreement one of the objects with one consciousness, thereby means to distract him from another or others."[19] Here there is a perfect isomorphism with what happens in the world of quantum phenomena, where, for example, light behaves experimentally *either* like waves *or* like corpuscles. But the results of a scientific experiment are acquired, by definition, on our own scale, in an inevitably classical world that is unable to conceive the unity of opposites. This separation between opposites (or contradictories) is due to our logic, to our language, to our way of interpreting the results on an infinitely more *complex* scale than the quantum one. On a quantum level, the light is one: it is *both* wave and corpuscle.

The example of complementarity is very instructive, in particular for two reasons.

First of all, due to the imaginary, Bohr was able to capture the amazing parallel between the pathological functioning of the hysteric's consciousness and the interpretation of quantum phenomena based on classical logic. Bohr felt that the source of structural schizophrenia in the classical vision lies in the complete separation of subject and object. Bohr wrote: "However, I hope that the concept of complementarity is able to explain current difficulties that have a profound analogy with general difficulties in the formation of human concepts resulting from the need to make a distinction between subject and object."[20]

At a deeper level, the example of complementarity shows the existence of general laws that are manifested in different forms at different scales. Far from leading to vague analogies, the idea of the unity of contradictories that underlies the principle of complementarity acquires the strength and precision of a symbol. Only the imaginary is able to embrace the infinite richness of a symbol. In this way, it can be understood why one of Bohr's favorite adages was Schiller's distich: "Only wholeness leads to clarity, and Truth lies in the abyss."[21]

UNDERSTANDING THE REALITY OF THE IMAGINARY: THE IMAGINARY AND THE IMAGINAL

The link between the expression of the imaginary in scientific research and in other fields of knowledge (artistic or religious, for example) has not yet been much explored.

An old problem (which interested Kepler, among others) is thus repeated: does any reality have a mathematical structure? The mathematical nature of certain visions of certain poets such as Henri Michaux[22] and René Daumal[23] surprises and intrigues.

One thing seems certain: this is what might be called *the reality of the imaginary*. In scientific creation, there is a natural shield against hallucinations, against delirium (whatever its apparent consistency), and against the dreaming state: information permeated by the imaginary is then submitted to the test of logical demonstration, in a well-established formal framework, and to the test of scientific experimentation. The history of science is littered with examples of results obtained long before the elaboration of demonstration methods (the cases of Fermat, Galois, Riemann, and Poincaré are among the best known). Mathematics and modern theoretical physics abound with such examples of abstract, formal results that were long considered mathematical curiosities without any connection to reality but that consequently found successful physical applications.

Moreover, as in most cases studied in this chapter, we should not use the term *imaginary* but rather that of *imaginal*, introduced by Henry Corbin[24] to denote the true, creative, visionary, vital, essential, founding imaginary: without vision, the real dissolves in an endless chain of veiled, distorted, mutilated images. "Following Corbin," Gilbert Durand wrote, "one may say that the *cogito* of the Imaginary is 'something that looks at me because it is in the very act of looking'—and I can only testify and profess my own vision, that the understanding of the one who looks, of the object looked at, and of the vision that allows such 'regard' are completed."[25]

The compliance of the human mind and of creativity in science and art is an open field, whose systematic exploration would allow the elucidation of numerous issues regarding the relationship between the real and the imaginary and the purpose of knowledge.

CHAPTER SIXTEEN

CAN SCIENCE BE A RELIGION?

> When you make from two one, you will become the son of man, and when you say: Mountain, move away! it will move.
>
> —*The Gospel of Thomas*, logion 106

Our instinct, our culture, and even our common sense will tell us that a positive answer to the question asked in the title of this chapter is simply absurd. The problem stated would therefore be a false problem. And yet, who could deny that modern science shattered and swept away the myths and beliefs that guided people's lives for centuries? And who could deny that modern science, by its most visible consequence—technology—is going to deeply disturb our lives and leave us disarmed when facing the dilemma of choosing between external wealth and the depletion (up to annihilation) of our inner life?

Science is in the situation of someone who is called to clean an apparently very beautiful house that is in horrible condition. Once the job is done, the person notices that the householder is often missing from home and the house is empty. Wouldn't this person be tempted to take possession of this empty place?

THE CLOWNS OF THE IMPOSSIBLE

Basic science sinks its roots into the nourishing land of interrogations shared by the whole area of human knowledge: what is the meaning of life? What is the role of the human being in the cosmic process? What is the role of nature in knowledge? Therefore, fundamental science has the same roots as religion, art, and mythology. Gradually, these investigations came to be considered ever less scientific and have been banished to the hell of the irrational, an area reserved for the poet, the mystic, and the philosopher. One important reason for this paradigm shift is represented by the indisputable triumph, in terms of direct materiality, of analytical, mechanistic, and reductionist thinking. It was enough to postulate laws that came from nowhere. Under those laws, under those equations of motion, everything could be predicted with precision once the initial conditions were known. Therefore, everything was determined, even predetermined. The *God hypothesis* was no longer necessary. The distance between "the Lord God," tolerated at best as a starting point, and this world's affairs became impassable. In this universe of false freedom (false because everything was predetermined), it was amazing that anything could truly happen. Witness to an absolute, static, and immutable order, scientists could no longer be, as formerly, philosophers of nature—they were forced to become *technicians of the quantitative*.

The establishment of quantum physics in the early twentieth century showed the fragility of such a paradigm. Quantum physics proved the unfounded nature of blind faith in continuity, in local causality, in mechanistic determinism (see chapter 4). Discontinuity was entering the royal gate—that of scientific experience. Local causality was making room for a finer concept, that of global causality, while the frightened reductionists were living the nightmare of the renaissance of the old concept of finality. The object was replaced by the *relationship, the interaction, the interconnection* of natural phenomena. This is one of the basic features of our new era—that of cosmodernity.

Finally, the classical concept of matter was replaced by the infinitely more subtle concept of substance/energy/space-time/information. The omnipotence of substance, the touchstone of the reductionists of all times, was shattered. Substance was simply one of the possible facets of matter.

With Planck and Einstein, an unprecedented revolution began that logically should have led to a new system of values designed to govern our lives. However, a century after the emergence of the quantum image of the world, nothing has really changed. We continue to act, consciously or unconsciously, according to the old concepts of previous centuries. Where does this schizophrenia between a quantum universe and a person who bears the moral authority of an outdated image of the world come from? Where does this disregard for the fundamental problems, left inside the locker room of ridiculous luxury, come from? Why are we helplessly witnessing the distressing spectacle of an ever more accelerated fragmentation, of a self-destruction that does not dare to pronounce its name? Why is the wisdom of natural systems ignored and obscured? Have we become *clowns of the impossible*, manipulated by an irrational force that we ourselves began?

HIGHLIGHTS OF THE NEW BARBARITY

There is a logical basis of fragmentation. Our physical body aims to assume its own identity, its own permanence. In its fascination with survival, it stubbornly fights to achieve noncontradiction through an ever-increasing *homogenization* of its surrounding environment. Its best tool during this struggle is the brain, especially due to its mechanical and automated functioning, which we will call in what follows *the mental*. The mind is a wonderful tool for differentiation, for distinction, for the detection of environmental hazards. It could therefore fully fulfill its role of adapting the body to the external environment. But in order to assume this task, it would need a link. The only link that can be developed is the recognition of humanity's place in the cosmic process, of its role in this process. If this role is ignored or forgotten, there is a reversal; the mind starts to act on its own account. In this way, the human being begins to conquer nature, instead of meeting the need for harmonious cooperation with it. A chaotic and anarchic proliferation of the mind inevitably invades the world. The whole world becomes like the physical human body, like its sense organs. This destructive movement of nature necessarily accelerates the mechanization, fragmentation, and annihilation of interaction.

So why should we marvel at the improper reign of quantity, analyzed so astutely by René Guénon?[1] Why should we marvel any longer at the idolatry of the computer as an almighty god in the pathetic imitation of some forgotten spirituality? The mental is, by its very nature, the chief imitator. Science wants to imitate religion, and religion wants to imitate science. The confrontation between scientists' imperialism and mystical imperialism does nothing but accelerate the fragmentation of terrestrial life. Is the current approximation between science and religion a sign of weakness or a sign of strength? It is not, of course, about denying the indisputable, intrinsic value of technoscience, which could have a place in the harmonious development of humanity. What is at issue here is the anarchical proliferation of technoscience, which finds its peak in the fact that the means of destruction on our planet are sufficient to destroy the Earth several times over.

As Michel Henry writes, the signs of the new barbarity are discernible everywhere in the world.[2] The origin of the new barbarity seems to lie in the explosive mixture between the binary thinking of the excluded third, a pure product of the mind, and a technology without any humanist perspective.

BETWEEN THE ANECDOTE AND THE UNSPEAKABLE

Does nature have something to tell us about ourselves? Is it true that by knowing the universe I can know myself? Are the two levels of knowledge corresponding to our double nature irremediably separated, the transition being entirely discontinuous? Then where arises the confidence in an *isomorphism* between different planes of knowledge that is reinforced daily by the progress in the study of physical laws?

Besides, what does "to know yourself" mean—obsessive though absurd imperative in terms of common logic—because how could a complex system fully decode another system of equal complexity? Must we open ourselves to a level of another complexity? And is this "openness" merely a word that hides the inaccessibility of the unknown?

Finally, do all these questions really mean anything to science as it is defined today? Should we not content ourselves with considering science to be a set of operative recipes in terms of direct materiality but with no significance in terms of being? Shall we accept the *how* but forget the *why*—to chase being outside the scope of science and fall again, this way, into an empty, divided world, without any sign, forgetting the warning of *The Gospel of Thomas*: "He who has recognized the world has found the body, but he who has found the body is superior to the world."[3]

How can we not marvel at the attempt of those who wish to see the spirit or God everywhere but use scientific data abusively? Science gives us a picture of an ever-changing reality. There is nothing stable or permanent in the claims of science. The only advance is that of interrogation. But how, also, can we not marvel at the position of those who, ignoring scientific data, proclaim the existence of an unbridgeable separation between science and religion?

The creative act of science, which results from the harmony between humanity and nature, is not just a mental act. It also includes intuition, through openness toward the kingdom of the subconscious (which we should rather call the superconscious). The creative act of science cannot be reduced to simple empirical evidence. It happens in a rather abstract, mathematical world, far from any empirical evidence. Mathematics, the real science of the analogy between analogies, becomes an organic part of scientific construction, the point that allows the existence of an isomorphism between the various sciences. How could one not see in this mathematical structure the very essence of the harmony between human and nature?

Are we, then, going to rediscover the old faith in the absolute objectivity of science, without any intervention by the subject? Certainly not. The quantum universe, the universe of interconnectedness, of nonseparability, implies a *participation* of the Subject, a true microcosmos reflecting the macrocosmos. It is precisely this participation, this harmony, that tells us that it is impossible to talk seriously about an absolute objectivity or about an absolute subjectivity of science. This is another basic feature of our new era—that of cosmodernity.

For this reason, can we draw the extreme conclusion of a social construction of scientific laws?

The falsity of such a conclusion consists in the refusal to distinguish between science and the scientist. Scientists themselves are the victim of modern schizophrenia: classical tools in their everyday life, through their usual language, through their conduct, they are also quantum instruments in their dialogue with nature, where they are constructed in mathematical language. *Faced with the unspoken, they transform the unknown into an anecdote.* Science follows its own path of progressive development of the harmony between humanity and nature at any cost.

Thus, a flurry of questions is imposed on us.

What is objectivity in relation to intersubjectivity? What is objectivity in the logic of the included third? Does the outside world remain the only possible reference point for objectivity? How can we avoid getting lost in the labyrinth of the inner life? How can we avoid falling once again into psychology, vulgarity? What is the sign of an objective event? Can the inner life be a measure of the interaction between different systems of systems? How can we avoid falling into absolute negation? How can we distinguish objectivity from certainty whose psychological foundation no longer needs to be proven? Is objectivity linked to transformation while *simultaneously* acting on the inner and the outer world? How can we overcome the approximation of the measurable through, at the same time, external action and romantic delusion? How can we avoid, at the same time, mystical romanticism and the vanity of scientism?

In any case, a *new objectivity* appears to arise from contemporary science, an objectivity that is no longer linked only to an object but to the subject/object interaction. New concepts should be invented. In this way, one would be able to speak about the *subjective objectivity* of science and the *objective subjectivity* of Tradition. Our potential, our opportunity, is that we can make the two poles of a fertile contradiction coexist in ourselves simultaneously.

THE SOKAL AFFAIR: BEYOND THREE EXTREMISMS

The Sokal affair started with a hoax. In 1994, a mathematical physicist from the New York University who was unknown outside a closed circle of physicists sent an article to the journal *Social Text* entitled "Transgressing the Boundaries: Toward a Transformative Hermeneutics of Quantum Gravity."[4]

The text was peppered with accurate quotations by physicists such as Bohr and Heisenberg and by philosophers, sociologists, historians of science, and psychoanalysts such as Kuhn, Feyerabend, Latour, Lacan, Deleuze, Guattari, Derrida, Lyotard, Serres, and Virilio. The bibliography also listed authors such as Lupasco, the included middle being given in the text as an example of "feminist logic" . . . However, Sokal's commentary made a number of absurd assertions, creating the impression that he was perfectly attuned to postmodernism and in particular to the *Cultural Studies* current of thinking. The journal's editors were delighted with the apparent adhesion of a physicist to their cause and published Sokal's text without hesitation, without even the slightest verification.

Almost as soon as the article appeared in print, Sokal himself revealed the hoax in a further article published in Lingua Franca[5] and, from that moment on, buoyed by the effect of the Internet, his fame was guaranteed. On a political level, Sokal wanted to show his friends in the American left wing that a revolution or social transformation could not be carried out based on the notion of reality as presented by philosophical relativism. Only physics (according to Sokal's view of what physics is) could play such a role.

This unleashed a storm of reactions on the Internet and the publication of books and countless journal articles on the exposure of a very real problem. For some, Sokal was the apostle of enlightenment against postmodern obscurantism. For others, he was part of a philistine imposture.

The Sokal Affair had the merit of highlighting a phenomenon that is becoming increasingly pervasive in contemporary culture: that of the radicalization of relativism. Indeed, it is evident that *relativist extremism* has a nefarious capacity to appropriate the language of the exact sciences. Taken out of context, this language can then be manipulated to say almost anything—indeed, to demonstrate that everything is relative. The first victims of this form of deconstruction are the exact sciences, which find themselves relegated to the rank of mere social constructs among many others, once the constraint of experimental verification is put into parentheses. It is, therefore, not surprising that in a few months Alan Sokal became the hero of a community that is aware of a blatant contradiction between its everyday practices and its social and cultural representation.

Paradoxically, however, the Sokal Affair served to reveal the extent of another form of extremism: namely, *scientist extremism*, the mirror image of *religious extremism*. Indeed, Sokal's position was supported by some notable heavyweights, including the Nobel Prize winner Steven Weinberg, who wrote a long article about it in the *New York Review of Books*.[6]

For Weinberg, "The gulf of misunderstanding between scientists and other intellectuals seems to be at least as wide as when C. P. Snow worried about it three decades ago." But what is the cause of this "gulf of misunderstanding"? *According to Weinberg, one of the essential conditions for the birth of modern science was the separation of the world of physics from the world of culture*. Consequently, from that moment on, any interaction between science and culture could only be seen as detrimental. In one fell swoop, with this he dismissed the philosophical considerations of the founding fathers of quantum mechanics as irrelevant.

Weinberg's arguments caused some to smart as if they scented the scientism of another century: the appeal to common sense in support of the claim about the reality of physical laws, the discovery through physics of the world "as it is," the one-to-one correspondence between the laws of physics and "objective reality," the hegemony on an intellectual level of natural science (because "we have a clear idea of what it means for a theory to be true or false"). However, Weinberg is certainly neither a positivist nor a mechanist. He is without doubt one of the most distinguished physicists of the twentieth century and a man of broad culture, and due attention should be paid to his arguments.

Weinberg's central idea, repeatedly pounded out like a mantra, concerns the discovery by physics of the existence of *impersonal laws*, the impersonal and eternal laws that guarantee the objective progress of science and that explain the unfathomable gulf between science and culture. The tone of the argument is overtly prophetic, as if in the name of a strange religion without God. One is almost tempted to believe in the immaculate conception of science. We are led to understand that for Weinberg, what is really at stake in the Sokal Affair is the *status of truth and reality*. According to Weinberg, truth, by definition, cannot depend on the social environment of the scientist. Science is the custodian of truth and, as such, its severance with culture is complete and definitive. There is only one Reality: the objective reality of physics. Weinberg states unequivocally that for culture and philosophy, the difference between quantum mechanics and classical mechanics or between Newton and Einstein's theories of gravitation is insignificant, a perspective in contradiction with the influence that these theories had on art (see chapter 14). The contempt with which Weinberg treats the notion of hermeneutics therefore seems quite natural.

Weinberg's conclusion comes down like a cleaver: "the conclusions of physics may become relevant to philosophy and culture when we learn the origin of the universe or the final laws of nature"; that is to say, never!

In 1997, Sokal decided to write a counterargument to the famous *Social Text* article, to express his actual thinking. For this he took on as a coauthor the Belgian physicist Jean Bricmont, who was supposed to have a good knowledge of the intellectual situation in France. The collaboration resulted in the publication of *Intellectual Impostures* in France, a book that, via its provoking title, was intended to become a best seller.[7] The content of the book surprised readers with its intellectual patchiness, and the repercussions were far less than those from the resounding success of Sokal's original hoax. The impostors in question—most of whom were French—were Jacques Lacan, Julia Kristeva, Luce Irigaray, Bruno Latour, Jean Baudrillard, Gilles Deleuze, Felix Guattari, and Paul Virilio.

It would be tedious to rehash here the nature of the "imposture." The method of the American and Belgian physicists was simple: sentences were removed from their original context and then shown to be meaningless or inaccurate from the perspective of mathematics or physics. For example, Lacan wrote that "In this pleasure space (*espace de jouissance*), to take something that is bounded, closed, constitutes a locus; to speak of it is a topology." Sokal and Bricmont's commentary follows: "In this sentence, Lacan uses four mathematical technical terms (*space, bounded, closed*, and *topology*), but without paying attention to their *meaning*; this sentence is meaningless from a mathematical point of view."[8] The method used by the authors discredits the book, and people thought that the Sokal Affair was definitively closed.

But Sokal came back. In 2005, he published a new book, again in France, with the catchy title *Pseudoscience and Postmodernism: Antagonists or Fellow-Travelers?*[9] We note in passing that Jean Bricmont wrote or contributed to a third of the book. Indeed, in the introduction, he sets the scene by quoting Jerry Fodor in the first lines: "The point of view to which I adhere . . . is *scientism*."

It is undoubtedly the case that drawing comparisons between postmodernism and pseudoscience has a certain appeal, even if Sokal concedes that he has found only a few postmodern thinkers who rally to the cause of pseudoscience.

The real novelty of the book is elsewhere, however, and can be found in Appendix A, which is entitled "Religion as Pseudoscience." It should be made clear here that Sokal is not referring to cults or new religious movements but to established religions—Christianity, Judaism, Islam, and Hinduism. It is, therefore, hardly surprising that Sokal points an accusatory finger at Pope John Paul II, described by Sokal as the head of a "major pseudoscientific cult," Catholicism.[10]

This last affirmation smacks simply of defamation, although it is interesting, nonetheless, to try to understand why religion for Sokal (and Bricmont) is a pseudoscience in the same way that astrology is.

Sokal candidly explains that "religion refers to real or alleged phenomena, or real or alleged causal relations, which modern science views as improbable." He goes on to say "that religion attempts to base its affirmations on a form of reasoning and system of proofs that far from satisfies the criteria of modern science in terms of logic and verification."[11] Sokal's epistemological error is plain to see: he assumes that modern science is the only arbiter of truth and Reality. At no point does he entertain the idea of a plurality of levels of Reality, with science applying to certain levels and religion to others. For Sokal, there can be only one level of Reality, a hypothesis that is epistemologically untenable in light of what modern science has taught us.

Sokal's attitude is strangely reminiscent of Lenin's position in 1908, in *Materialism and Empiriocriticism*, where he attacked theories of physics that implied multidimensional space-time by proclaiming that revolution could only be carried out in four dimensions. Like Sokal, Lenin believed in the existence of a single level of Reality. Or rather, he stuck to this belief to justify his revolution.

The three extremisms at stake here, of which religious extremism is the one with which the public is most familiar, have brought to the fore a fundamental problem—that of the status of truth and Reality—and have shown us the consequences, including the political consequences, of this problem.

What one needs to do at present is to go beyond these three extremisms, which are the germ of new forms of totalitarian systems. The Sokal affair gives us a good opportunity to reformulate not only the conditions for a dialogue between the hard sciences and the humanities but also a dialogue between science and culture, science and society, science and religion, and science and spirituality on a new and rigorous basis. Ultimately, the source of the violent polemic unleashed by the Sokal Affair is the considerable confusion between the *tools* and the *conditions* of the dialogue. Sokal and his friends are right to denounce the anarchic migration of concepts from the hard sciences to the humanities, which can only lead to a semblance of rigor and validity. But moderate relativists are also right to denounce the desire expressed by certain scientists to ban the dialogue between science and culture. Indeed, Weinberg falls prey to exactly the same form of confusion as his critics. Why does he declare himself to be "against philosophy" (the title of one of the chapters of his book *Dreams of a Final Theory*)?[12] Simply because he criticizes the tools of philosophy as not being productive in the process of scientific creation.

If there is a dialogue between the different disciplines, it cannot be founded on the concepts of one or another discipline but only on what the disciplines have in common: the Subject itself, in interaction with the Object, which refuses any form of formalization and always maintains an element of irreducible mystery. Throughout the twentieth century, the Subject has been considered as the Object: an object of experience, an object of ideologies that claimed to be scientific, an object of "scientific" studies aimed at dissecting it, formalizing and manipulating it, and revealing, in so doing, a self-destructing process in the irrational struggle of human beings against themselves. At the end of the day, the Sokal Affair sends us to the resurrection of the Subject: a veritable transdisciplinary quest, the painstaking working out of a new art of living and thinking.

A NECESSARY ISOMORPHISM

Systemic and quantum thinking, which are based on quantum physics but go beyond the narrowness of science, conceive of the universe as a vast whole, a vast cosmic matrix, where everything is in perpetual movement and energetic structuring. The real movement is the movement of energy. Objects are but local configurations of matter. But this unity of the world is not static; it implies differentiation, diversity, contradiction. The world is in a state of eternal genesis. Cosmodernity is also that—the recognition of a vast cosmic matrix to which we belong.

The separation between spirit and matter can be considered a mental construct based on the classical image of Reality. Modern science has actually become the kingdom of the invisible, of the organization of matter at ever more subtle levels of materiality.

Assuming that the vast whole is not itself a pure mental construct, how can we tackle it, how can we understand it? Overwhelmed by the crowd of apparently separate phenomena on our scale, it is impossible for us to undo everything starting from this fragmentation. The only possibility that remains for us is to understand the laws of the division of the whole.

Even if one day we do understand the laws of the division of the whole in natural systems, how can this clarify the laws of our inner life? Is there an isomorphism between the supersensible world and the world of events, of the sense organs, as was believed, for instance, by Dante? In *The Divine Comedy*, Dante sees God as a point of intense brilliance. Around this source-point, nine circles are spinning, animated by an increasingly faster motion as the circles become closer to the source-point. In contrast, the nine spheres spinning around the Earth are animated by an increasingly faster rotation as the circles become more distant. The sphere most distant from Earth corresponds to the circle nearest to the source-point.[13]

THE END OF SCIENCE?

After God's death, humanity's death, the death of nature, and the end of history, are we going to experience the end of science? A best seller in the United States has sparked a huge scandal on this topic.[14] The book, *The End of Science*, was denounced as dangerous by the scientific adviser of former President Clinton, by the administrator of NASA, and by many Nobel Prize winners, not to mention many critics from different parts of the world. What is the basis of this scandal?

John Horgan is a journalist at *Scientific American*, and his competence is not questioned. What is questioned, though, is his thesis on scientific knowledge. Horgan based his work on the finding that everything in this world *has a limited lifetime*. A human being lives for a finite amount of time; the particle lives for another finite amount of time; the universe itself may be mortal. Like empires, cultural models also have a lifetime. Therefore, we could ask whether modern scientific knowledge, born in the sixteenth century, has already lived its life. More specifically, we could ask if, after three great intellectual constructions of the twentieth century—the theory of relativity, quantum physics, and the discovery of the genetic code—scientific knowledge has not reached its limits. Of course, this is not about the end of science as a human activity—scientists will have detailed work until the end of time—it is about the end of what the human spirit can discover of the fundamental laws of the universe. As shown, Horgan's question is less naive than the title of the book might allow us to assume. Some of the most brilliant scientists of our time were questioned by Horgan on their views about the possible end of science.

The answers cover an extremely wide spectrum, from an unconditional "yes" to a violent "no." The great value of the book consists in the fact that, in the face of such a general matter, with such important consequences, the people interviewed are forced to flee the wooden language that very often characterizes, alas! the dialogue between scientists and journalists and to submit instead their private beliefs. There are many surprises.

Thus, when John Horgan asked Karl Popper whether his famous hypothesis regarding the falsification of theories is itself falsifiable, Popper replied: "This is one of the most idiotic criticisms that one can imagine!" When asked about the absence of any testable experimental predictions of the superstring theory, Edward Witten replied that this theory predicts gravity, although he admitted that the greatest possibility of the theory's truth consists in its self-consistency, its elegance, its beauty. Horgan is pretty severe, when you consider that, in fact, Witten, the most brilliant of all contemporary physicists, is a practitioner of the "ironic science," which is related more to literary criticism or to philosophy. The most surprising remark comes from Murray Gell-Mann, who discovered quarks, when he said that the problem of consciousness is the last refuge of obscurantism and mystification. In a more abrupt way, Steven Weinberg went in the same direction: for him, science rules out any divine plan, and the grand unified theory of physics will eliminate the mystical spirit.

At the other end of the spectrum, it is refreshing to read the balanced replies of Ilya Prigogine and Freeman Dyson. Expressing his belief in the rebirth of physics, Prigogine said: "It is not possible to believe that we are a part of an automaton if we continue to believe in humanism." As for Dyson, he does not exclude the end of physics as we know it today: "The end of physics could be the beginning of something else."

THE SPIRITUAL DIMENSION OF DEMOCRACY: UTOPIA OR NECESSITY?

There is a great spiritual poverty present on our Earth. It manifests as fear, violence, hatred, and dogmatism. In a world with more than eight thousand academic disciplines, more than ten thousand religions and religious movements, and more than six thousand tongues, it is difficult to dream about mutual understanding and peace. There is an obvious need for a new spirituality that will conciliate technoscience and wisdom. Of course, there are already several different kinds of spirituality that have been present on our Earth for centuries and even millennia. One might ask: why is there a need for a new spirituality if we have them all here and now?

Before answering this question, we must face a preliminary question: is a big picture still possible in our postmodern times?[15] Radical relativism answers this question in a negative way. However, its arguments are not solid and logical. For radical relativists, after the death of God, the death of humanity, the end of ideologies, the end of history (and perhaps tomorrow, the end of science and the end of religion), a big picture is no longer possible. For cosmodernity, a big picture is not only possible but also vitally necessary, even if it will never be formulated as a closed theory. Fifty years ago, the great quantum physicist Wolfgang Pauli wrote: "Facing the rigorous division from the seventeenth century of the human spirit in isolated disciplines, I consider the aim of transgressing their opposition . . . as the explicit or implicit myth of our present times."[16]

The first motivation for a new spirituality is technoscience, associated with fabulous economic power, which is simply incompatible with present spiritualities. It drives a hugely irrational force of efficiency for efficiency's sake: everything that can be done will be done, for better or worse. The second motivation for a new spirituality is the difficulty of dialogue between different spiritualities, which often appear as antagonistic, as one can testify to in our everyday life.

Simply put, we need to find *a spiritual dimension of democracy*. Social and political life go well beyond academic disciplines, but they are based upon the knowledge generated by these disciplines. We therefore need transdisciplinarity. Transdisciplinarity can help with this important advancement of democracy through its basic notions of *transcultural* and *transreligious*[17] (see chapter 1).

Homo religiosus probably existed from the beginnings of the human species, at the moment when human beings tried to understand the meaning of our life. The *sacred* is our natural realm. We tried to capture the unseen from our observation of the visible world. Our language is that of the imaginary trying to penetrate higher levels of Reality—parables, symbols, myths, legends, revelation.

Homo economicus is a creation of modernity. We believe only in what is seen, observed, measured. The *profane* is our natural realm. Our language is that of just one level of Reality, accessible through the analytic mind—hard and soft sciences, technology, theories and ideologies, mathematics, informatics.

The only way to avoid the dead end of the homo religiosus versus homo economicus debate is to adopt *transdisciplinary hermeneutics*.[18] Transdisciplinary hermeneutics is a natural outcome of transdisciplinary methodology.

The transdisciplinary approach to Reality allows us to define three types of meaning:

1. *Horizontal meaning*—interconnections at one single level of Reality. This is what most of the academic disciplines do.

2. *Vertical meaning*—interconnections involving several levels of Reality. This is what poetry, art, and quantum physics do.

3. *Meaning of meaning*—interconnections involving all of Reality: the Subject, the Object and the Hidden Third. This is the ultimate aim of transdisciplinary research.

Cultures and religions are not concerned, as academic disciplines are, with fragments of levels of Reality only: they simultaneously involve one or several levels of Reality of the Object, one or several levels of Reality of the Subject, *and* the nonresistance zone of the Hidden Third. Technoscience is entirely situated in the zone of the Object, but cultures and religions cross all three terms: the Object, the Subject and the Hidden Third. This asymmetry demonstrates the difficulty of dialogue between technoscience and culture or religion: such a dialogue can occur only when there is a *conversion* of technoscience to values, that is, when the technoscientific culture becomes a true culture.[19] It is precisely this conversion that transdisciplinarity is able to perform. This dialogue is methodologically possible, because the Hidden Third crosses all levels of Reality.

Technoscience is in a quite paradoxical situation. In itself, it is blind to values. However, when it enters into dialogue with cultures and religions, it becomes the best mediator for the reconciliation of different cultures and different religions.

Transdisciplinary hermeneutics is able to identify the common germ of homo religiosus and of homo economicus, which can be called *homo sui transcendentalis*.

In the more or less long term, one can predict that transdisciplinary hermeneutics will lead to what Hans-Georg Gadamer calls a *fusion of horizons*,[20] not only of science and religion but also of all other fields of knowledge, such as arts, poetry, economics, social sciences, and politics, so crucial in the science/religion debate. Transdisciplinary hermeneutics avoids the trap of trying to formulate a superscience or a superreligion. Unity of knowledge can be only an open, complex, and plural unity.

Homo sui transcendentalis is in the process of being born. Each of us will not be some new person but a person reborn. This new birth is a potentiality inscribed in our very being.

Our language is generated by the notions of levels of Reality of the Subject, of the Object, and of the Hidden Third. In transdisciplinary hermeneutics, the classic real/imaginary dichotomy disappears. We can think of a level of Reality of the Object or of the Subject as being a fold in the Hidden Third. The real is a fold in the imagination and the imagination is a fold in the real. The ancients were right: there is indeed an *imaginatio vera*, a foundational, true, creative, visionary imagination.

CHAPTER SEVENTEEN

THE HIDDEN THIRD AND THE MULTIPLE SPLENDOR OF BEING

PREMODERNITY, MODERNITY, POSTMODERNITY, AND COSMODERNITY AS DIFFERENT VISIONS OF THE RELATION BETWEEN THE SUBJECT AND THE OBJECT

The relation between the Subject and the Object is a crucial problem of philosophy.

This relation has varied in the different periods of human culture. In the premodern world, the Subject was immersed in the Object. In the modern world, the Subject and the Object were supposed to be totally separate, and in our postmodern era, the Subject has become predominant as compared with the Object (see Figures 17.1–17.3).

Of course, the key point in understanding the Subject/Object relation is the vision of Reality that humans shared in different periods of historical time.

Dictionaries tell us that *reality* means[1]: "1. the state or quality of being real. 2. resemblance to what is real. 3. a real thing or fact. 4. real things, facts, or events taken as a whole; state of affairs." These are clearly not definitions but descriptions in a vicious circle: "reality" is defined in terms of what is "real."

In order to avoid any ambiguity, *reality* is defined here in a sense that is used by scientists, namely in terms of *resistance* (see chapter 8).

In order to avoid further ambiguities, we have to distinguish the words *Real* and *Reality*. *Real* designates that which *is*; *Reality* is connected to resistance in our human experience. "The Real" is, by definition, veiled forever (it does not tolerate any further qualifications); but "Reality" is accessible to our knowledge. The Real involves nonresistance; Reality involves resistance.

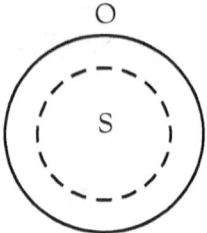

S = Subject, O = Object

Figure 17.1. The Relationship between Subject and Object in Premodernity.

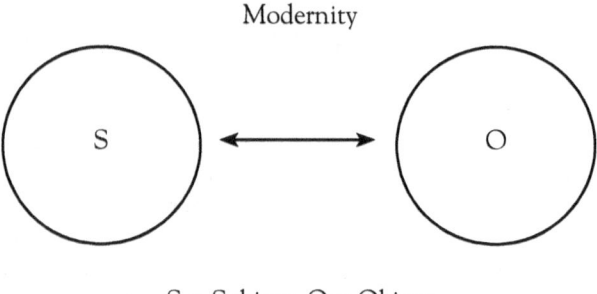

S = Subject, O = Object

Figure 17.2. The Relationship between Subject and Object in Modernity.

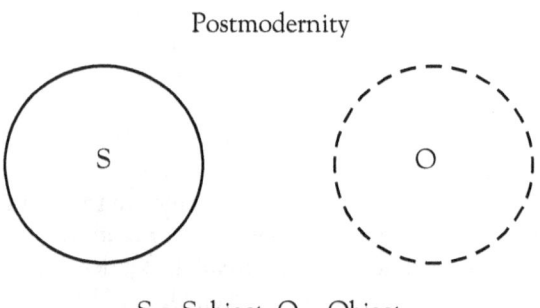

S = Subject, O = Object

Figure 17.3. The Relationship between Subject and Object in Postmodernity.

LADDER OF DIVINE ASCENT AND LEVELS OF BEING

In fact, the idea of "levels of Reality" is not completely new. From the beginnings of our existence, human beings have felt that there are at least two realms of reality—one visible, the other invisible.

Theological literature expressed the idea of a "scale of being" in a more elaborate way, which corresponds, of course, to a scale of Reality. The scale of Jacob (Genesis 28:10–12) is one famous example, agreeably illustrated in the Christian Orthodox iconography. There are several variants of the scale of being. The most famous one is found in the book *Climax* or *Ladder of Divine Ascent* of Saint John Climacus (ca. 525–606). The author, also known as John of the Ladder, was a monk at the monastery on Mount Sinai. There are thirty steps on the ladder, which symbolize the process of *theosis*. Resistance and nonresistance are well illustrated in the scale of John of the Ladder: the human being climbs the steps, which denote the effort of human beings being to evolve (from the spiritual point of view) through resistance to their habits and thoughts, and the angels, the messengers of God, help them to jump through the intervals of nonresistance between the steps of the ladder.

In the second part of the twentieth century, two important thinkers on the problem of levels of Reality were Nicolai Hartmann and Werner Heisenberg.

Nicolai Hartmann (1882–1950) is a somewhat forgotten philosopher, who had Hans-Georg Gadamer as a student and Martin Heidegger as his successor at the University of Marburg in Germany. He elaborated an ontology based on the theory of categories and distinguished four levels of Reality[2]: inorganic, organic, emotional, and intellectual. In 1940, he postulated four laws of the levels of Reality: the law of recurrence, the law of modification, the law of the *novum*, and the law of distance between levels.[3] The last law postulates that the different levels do not develop continuously but in leaps; it is therefore particularly interesting in the context of the contemporary view of Reality.

Almost simultaneously with Hartmann, in 1942, Werner Heisenberg, a Nobel laureate in physics, elaborated a very important model of levels of reality in his *Manuscript of 1942*,[4] which was only published in 1984.

The philosophical thinking of Heisenberg is structured by "two directory principles: the first one is that of the division of levels of Reality corresponding to different modes of objectivity that depend on the incidence of the knowledge process and the second one is that of the progressive erasure of the role played by ordinary concepts of space and time."[5]

For Heisenberg, reality is "the continuous fluctuation of experience as gathered by conscience. In this respect, it is never wholly identifiable to an isolated system."[6]

As Catherine Chevalley, who wrote the introduction to the French translation of Heisenberg's book, says: "for him, the semantic field of the word 'reality' included everything given to us by experience taken in its largest meaning, from experience of the world to that of the soul's modifications or of the autonomous signification of the symbols."[7]

In agreement with Husserl, Heidegger, Gadamer, and Cassirer (whom he knew personally), Heisenberg states often that one must suppress all rigid distinctions between Subject and Object. He also states that one must end with the privileged reference on the outer material world and that the only way to approach the sense of reality is to accept its division into regions and levels.

Heisenberg distinguishes between "regions of reality" (*der Bereich der Wirklichkeit*) and "levels of reality" (*die Schicht der Wirklichkeit*).

"By regions of reality," wrote Heisenberg, "we understand an ensemble of nomological connections. These regions are generated by groups of relations. They overlap, adjust, cross, always respecting the principle of noncontradiction." The regions of reality are, in fact, strictly equivalent to the levels of organization of systemic thinking.

Heisenberg is conscious that simple consideration of the existence of regions of reality is not satisfactory, because this will put classical and quantum mechanics on the same plane. It is for this essential reason that he regrouped these reality regions into different levels of Reality.

Heisenberg regrouped the numerous regions of reality into three distinct levels.

"It is clear," wrote Heisenberg, "that the ordering of the regions has to substitute the gross division of the world into a subjective reality and an objective one and that the world must stretch itself between these poles of subject and object in such a manner that at its inferior limit are the regions where we can completely objectify reality. In continuation, one has to join regions where the states of things could not be completely separated from the knowledge process during which we are identifying them. Finally, on the top, have to be the levels of Reality where the states of things are created only in relation with the knowledge process."[8]

The first level of Reality in the Heisenberg model corresponds to the states of things, which are objectified independently of the knowledge process. At this first level, he situates classical mechanics, electromagnetism, and the two relativity theories of Einstein; in other words, classical physics. The second level of Reality corresponds to the states of things inseparable from the knowledge process. Here he situates quantum mechanics, biology, and the consciousness sciences. Finally, the third level of Reality corresponds to the states of things created in relation to the knowledge process. On this level of Reality, he situates philosophy, art, politics, "God" metaphors, religious experience, and inspiration experience.

Note that religious experience and inspiration experience are difficult to assimilate into a level of Reality. They rather correspond to the passage between different levels of Reality, in the zone of nonresistance.

It is important to underline in this context that Heisenberg shows great respect for religion. In relation to the problem of God's existence, he wrote: "This belief is not at all an illusion but is only the conscious acceptance of a tension never realized in reality, tension that is objective and that advances in independently of the humans that we are and that is yet, in its turn, nothing but the content of our soul, transformed by our soul."[9]

The expression used by Heisenberg—"a tension never realized in reality"—is particularly significant. It evokes "Real" as distinct from "Reality."

For Heisenberg, the world and God are indissolubly linked: "this opening to the world that is at the same time the 'world of God' is, finally, the highest happiness that the world can offer us: the consciousness of being home."[10]

He remarks that the Middle Ages chose religion and the seventeenth century chose science, but today any choice or criteria for values has vanished.

Heisenberg also insists on the role of intuition: "Only intuitive thinking," wrote Heisenberg, "can pass over the abyss that exists between the system of concepts already known and that of new concepts; the formal deduction is helpless when it comes to throwing a bridge over this abyss."[11]

TOWARD A UNIFIED THEORY OF LEVELS OF REALITY

Transdisciplinarity is founded upon three axioms[12]:

1. The ontological axiom: There are different levels of Reality of the Subject and, correspondingly, different levels of Reality of the Object.

2. The logical axiom: The passage from one level of Reality to another is ensured by the logic of the included middle.

3. The epistemological axiom: The structure of the totality of levels of Reality appears, in our knowledge of nature, of society, and of ourselves, as a complex structure: every level is what it is because all the levels exist at the same time.

The introduction of levels of Reality induces a multidimensional and multireferential structure of Reality. Both the notions of the "Real" and "levels of Reality" relate to what is considered to be the "natural" and the "social" and are therefore applicable to the study of nature and society.[13]

Every level is characterized by its *incompleteness*: the laws governing this level are just a part of the totality of laws governing all levels. And even the totality of laws does not exhaust the entirety of Reality: we have also to consider the Subject and its interaction with the Object. *Knowledge is forever open*.

The zone between two different levels and beyond all levels is a zone of *nonresistance* to our experiences, representations, descriptions, images, and mathematical formulations (see chapter 17).

The unity of levels of Reality of the Object and its complementary zone of nonresistance constitutes what we call the *transdisciplinary Object*.

In agreement with the phenomenology of Edmund Husserl (1859–1938),[14] one asserts that the different levels of Reality of the Object are accessible to our knowledge thanks to the different levels of perception that are potentially present in our being. These levels of perception permit an increasingly general, unifying, encompassing vision of Reality without ever entirely exhausting it. In a rigorous way, these levels of perception are, in fact, *levels of Reality of the Subject*.

As in the case of levels of Reality of the Object, the coherence of levels of Reality of the Subject presupposes a zone of nonresistance to perception.

The unity of levels of levels of Reality of the Subject and this complementary zone of nonresistance constitutes what is called the *transdisciplinary Subject*.

The two zones of nonresistance of transdisciplinary Subject and Object must be identical for the transdisciplinary Subject to communicate with the transdisciplinary Object. A flow of consciousness that coherently cuts across different levels of Reality of the Subject must correspond to the flow of information coherently cutting across different levels of Reality of the Object. The two flows are interrelated because they share the same zone of nonresistance.

Knowledge is neither exterior nor interior: it is simultaneously exterior and interior. Studies of the universe and of the human being sustain one another.

The zone of nonresistance plays the role of a *third* between the Subject and the Object, an interaction term that allows the unification of the transdisciplinary Subject and the transdisciplinary Object but preserves their difference. In the following, this interaction term is called the *Hidden Third*.

There is a big difference between the Hidden Third and the included third: *the Hidden Third is a-logical*, because it is entirely located in the area of nonresistance, while *the included third is logical*, because it refers to the contradictories A and non-A located in the area of resistance. But there is also one similarity. Both of them unite contradictories: A and non-A in the case of the included third and Subject and Object in the case of the Hidden Third. *The Subject and the Object are the supreme contradictories*: they not only cross the area of resistance but also that of nonresistance. Thus, it is understandable why, in the view of some Christian thinkers, such as Jacob Boehme, when God decides to create the world (and thus to know himself), he places the contradiction at the origin of the world. It is understandable also why the Hidden Third is the one that gives meaning to the included third, because in order to unite the contradictories A and non-A located in the area of resistance, the included third must cross the area of nonresistance: the included third is actually a "middle-without-name." This is precisely wherein lies the great difficulty of formulating a true logic of the included middle, which must necessarily integrate the discontinuous leap between the levels of Reality. This new logic will be a *transcategorical* one. If compatibility between the levels of Reality and the included third is certain, however, their reconnection inside certain logic will not be achievable according to the patterns of known logics. Despite efforts made so far, the problem remains open.[15]

The role of the Hidden Third and of the included middle in the transdisciplinary approach to Reality is, after all, not so surprising. The words *three* and the prefix *trans* have the same etymological root: "three" means "the transgression of two, what goes beyond two." Transdisciplinarity means transgression of duality, opposing binary pairs: subject-object, subjectivity-objectivity, matter-consciousness, nature-divinity, simplicity-complexity, reductionism-holism, diversity-unity. This duality is transcended by the open unity that encompasses both the universe and the human being.

The Hidden Third, in its relationship with the levels of Reality, is fundamental for the understanding of *unus mundus* described by cosmodernity. Reality is simultaneously a single and a multiple One. If one remains confined to the Hidden Third, then the unity is undifferentiated, symmetric, situated in the *non-time*. If one remains confined to the levels of Reality, there are only differences, asymmetries, located in time. To simultaneously consider the levels of reality and the Hidden Third introduces a break in the symmetry of *unus mundus*. In fact, *the levels of Reality are generated precisely by this breaking of symmetry introduced by time*.

The ternary partition {Subject, Object, Hidden Third} is, of course, different from the binary partition {Subject versus Object} of classical, modern metaphysics.

Transdisciplinarity leads to a new understanding of the relation between Subject and Object, which is illustrated in Figure 17.4.

In the transdisciplinary approach, the Subject and the Object are immersed in the Hidden Third.

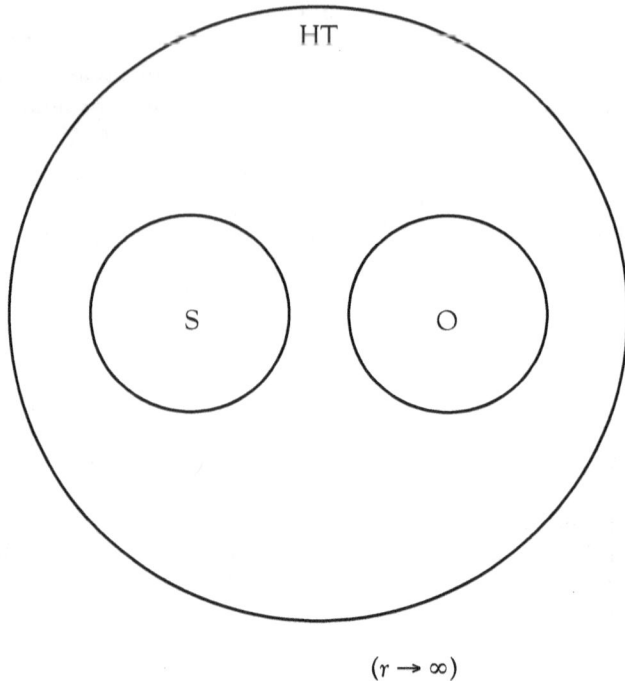

$(r \to \infty)$

S = Subject, O = Object, HT = Hidden Third

Figure 17.4. The Relationship between Subject and Object in Cosmodernity.

The transdisciplinary Subject and its levels, the transdisciplinary Object and its levels, and the Hidden Third define the transdisciplinary Reality, or *trans-Reality* (see Figure 17.5).

In Figure 17.5, the Hidden Third is constituted by the point X of contact between Object and Subject, the zone of nonresistance between the Object and the Subject, and the zone of nonresistance between the levels of Reality.

The incompleteness of the general laws governing a given level of Reality signifies that, at a given moment of time, one necessarily discovers contradictions in the theory describing the respective level: one has to assert A and non-A at the same time. It is the included third logic that allows us to jump from one level of Reality to another level of Reality (see chapter 10).

THE HIDDEN THIRD

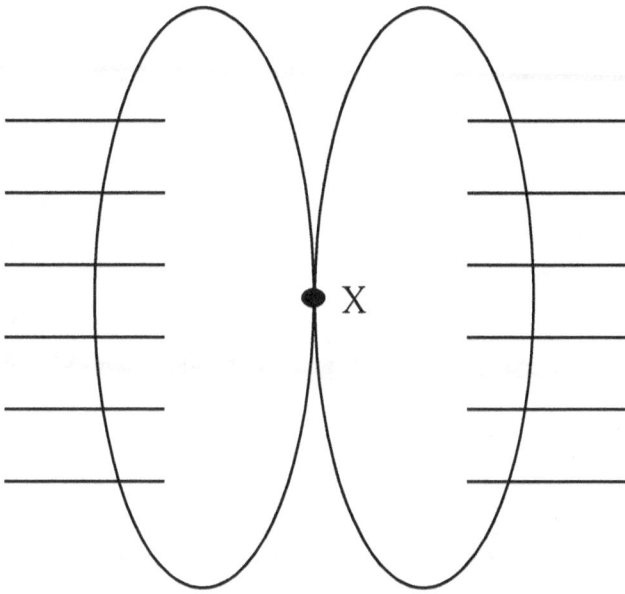

Figure 17.5. Trans-Reality.

All levels of Reality are interconnected through complexity. In fact, complexity is a modern form of the very ancient principle of universal interdependence. The principle of universal interdependence entails the maximum possible simplicity that the human mind can imagine, the simplicity of the interaction of all levels of reality. This simplicity cannot be captured by mathematical language, only by symbolic language.

The transdisciplinary theory of levels of Reality appears to reconcile reductionism with nonreductionism. It is, in some aspects, a multireductionist theory, via the existence of multiple, discontinuous levels of Reality. However, it is also a nonreductionist theory, via the Hidden Third, which restores the continuous interconnectedness of Reality. The reductionism/nonreductionism opposition is, in fact, a result of binary thinking, based upon the excluded middle logic. The transdisciplinary theory of levels of Reality allows us to define a new view on Reality, which can be called *transreductionism*.[16]

The transdisciplinary notion of levels of Reality is incompatible with a reduction of the spiritual level to the level of the psyche, of the level of the psyche to the biological level, and of the biological level to the physical level. These four levels are united through the Hidden Third; however, this unification cannot be described by a scientific theory. By definition, science excludes nonresistance. Science, as is defined today, is limited by its own methodology.

The transdisciplinary notion of levels of Reality leads also to a new vision of personhood, based upon the inclusion of the Hidden Third. The unification of the Subject is performed by the action of the Hidden Third, which transforms knowledge in *understanding*. As used here, *understanding* means fusion of knowledge and being.

In the transdisciplinary approach, the Hidden Third appears as the source of knowledge but, in its turn, needs the Subject in order to know the world: the Subject, the Object, and the Hidden Third are interrelated. This view is perfectly compatible with the theological approach developed by Christopher C. Knight in his book *The God of Nature: Incarnation and Contemporary Science*.[17]

The human person appears as an interface between the Hidden Third and the world. The erasing of the Hidden Third in knowledge signifies a one-dimensional human being who has been reduced to cells, neurons, quarks, and elementary particles.

This trans-Reality is the foundation of a new era—the cosmodern era. *Cosmodernity* means essentially that all entity (existence) in the universe is defined by its relation to all other entities. The human being, in turn, is related as a person to the Great Other, the Hidden Third. *The idea of cosmos is therefore resurrected.* This is the reason I introduced the word *cosmodernity* in 1994, in a book of aphorisms called *Poetical Theorems*.[18]

The present book gives the scientific and philosophical foundations of cosmodernity. The arguments coming from contemporary American literature, exposed in the book *Cosmodernism* by Christian Moraru,[19] are excellent and necessary complements.

In analyzing American narrative in the late-globalization era, Moraru identified several axes of his book: "These axes (a) thematize the cosmodern as a mode of thinking about the world and its culture, about cultural perception, self-perception, and identity; (b) bringing to the forefront, accordingly, the intersubjective-communicational, dynamic dimension of cosmodernism; and (c) articulating the cosmodern imaginary into five regimes of relatedness, or subimaginaries: the 'idiomatic,' the 'onomastic,' the 'translational,' the 'readerly,' and the 'metabolic.'"[20] The cosmodern mind is a "vehicle for a new togetherness for a solidarity across political, ethnic, racial, religious, and other boundaries."[21] A "new geometry of 'we'"[22] and a powerful *with-ness*[23] distinguish cosmodernity from modernity or postmodernity. All cultures are interrelated. Cosmodernity is, by its very nature, transcultural and transreligious. In agreement with what is said in the present book, Moraru asserts that "cosmodern rationality is relational. In cosmodernism, *relatio* is a new, sui generis *ratio mundi*."[24] Modern rationality is metamorphosed in relationality. Moraru coins the very evocative word "poethics,"[25] and he stresses that "cosmodernism is best understood as an ethical rather than 'technical' project. This project has considerable bearings on how we think, not just about the subject but also about discourse, history, culture, community, patrimony, and tradition."[26] The ethical imperative of cosmodernity is that of togetherness.[27] The entire world, our world, is a "web of ideas and images,"[28] of people, cultures, religions, and spiritualities.

Poets and writers perceive better than scientists all the potentialities of cosmodernity and of the Hidden Third. The great Spanish poet Clara Janés (b. 1940), who integrated the scientific vision of the world into her poetry,[29] wrote a wonderful poem entitled "The Hidden Third"[30]:

> To rest in the green
> of the forest,
> in the bird that calls out the alphabet,
> in the suspended drops of water,
> your letters
> beyond all concept
> descending on the foliage,
> like a gentle breath
> that tempers
> the dark swirling
> of the word.
>
> Return to me the virginal call
> in a form
> of pure resonance
> that pierces the heart
> and fills it with communicant light
> annulling the limits
> that establish the other
> through enunciation.
>
> And you, tired mouth,
> follow attentively
> the secret of the waves
> and learn
> transparency.

AT THE THRESHOLD OF NEW RENAISSANCE

The unified theory of levels of Reality presented in this book is valid in all fields of knowledge, which, at the beginning of the twenty-first century, involve more than eight thousand academic disciplines, every discipline claiming its own truths and having its own laws, norms, and terminology. The transdisciplinary theory of levels of Reality is a good starting point at which to erase the fragmentation of knowledge and, therefore, the fragmentation of the human being.

In this context, the dialogue of transdisciplinarity with apophatic thinking will be, of course, very useful. The Hidden Third is a basic apophatic feature of unified knowledge.[31] The dialogue with *biosemiotics*, as developed, for example, in the stimulating book *Signs of Meaning in the Universe*, by Jesper Hoffmeyer,[32] is also important. Biosemiotics is transdisciplinary by its very nature.[33] We live in the *semiosphere* as much we live in the atmosphere, hydrosphere, and biosphere. The human being is the only being in the universe that is able to conceive of an infinite wealth of possible worlds. These possible worlds certainly correspond to different levels of Reality. Powerful concepts elaborated by biosemioticians, such as *semiotic freedom*, could lead us to understand what "personhood" could mean.

"What is Reality?" asks Peirce.[34] He tells us that perhaps there is nothing at all that corresponds to Reality. It may be just a working assumption in our desperate tentative knowing. But if there is a Reality, says Peirce, it has to consist in the fact that *the world lives, moves, and has in itself a logic of events that corresponds to our reason*. Peirce's view of reason totally corresponds to the cosmodern view of Reality.

A unified theory of levels of Reality is crucial in building sustainable development and sustainable futures. The speculation on these matters up to now has been based upon reductionist and binary thinking: everything has been reduced to society, economy, and environment. The individual level of Reality, the spiritual level of Reality, and the cosmic level of Reality are completely ignored. Sustainable futures, so necessary for our survival, can only be based on a unified theory of levels of Reality. We are part of the ordered movement of Reality. Our freedom consists in entering into the movement or perturbing it. We can respond to the movement or impose our will on it with power and domination. Our responsibility is to build sustainable futures in agreement with the overall movement of Reality.

This book is about a new era—*cosmodernity*—founded on a new vision of the contemporary interaction between science, culture, spirituality, religion, and society. The old idea of the cosmos, in which we are active participants, is resurrected.

One idea goes through this book like an axis: *Reality is plastic*. Reality is not something outside or inside us: it is simultaneously outside and inside. We are part of this Reality, which changes due to our thoughts, feelings, and actions. This means that we are fully responsible for what Reality is. The world moves, lives, and offers itself to our knowledge thanks to ordered structures of something that is continually changing. Reality is therefore rational, but its rationality is multiple, structured on levels. It is the logic of the included middle that allows our reason to move from one level to another.

The levels of Reality correspond to the *levels of understanding*, in a fusion of knowledge and being. All levels of Reality are interwoven. The world is at the same time knowable and unknowable.

The Hidden Third between Subject and Object denies any rationalization. Therefore, Reality is also *transrational*. The Hidden Third conditions not only the flow of information between Subject and Object but also that between the different levels of Reality of the Subject and between the different levels of Reality of the Object. The discontinuity between the different levels is compensated by the continuity of information held by the Hidden Third. As the source of Reality, the Hidden Third reinforces perpetually its power of action through the interaction with levels of Reality in the transcosmic unity that includes us and the universe.

The irreducible mystery of the world coexists with the wonders discovered by reason. The unknown enters every pore of the known, but without the known, the unknown would be a hollow word. Every human being on this Earth recognizes his or her face in every other human being, independent of particular religious or philosophical beliefs, and all humanity recognizes itself in the infinite Otherness.

A new spirituality, free of dogmas, is already potentially present on our planet. The present book traces exemplary signs and arguments for its birth, from quantum physics to theater, literature, and art. We are at the threshold of a true New Renaissance, which asks for a new, cosmodern consciousness.

NOTES

CHAPTER ONE

1. Galileo Galilei, *Dialogue sur les deux grands systèmes du monde*, trans. René Fréreux and François de Gandt (Paris: Seuil, 1992).
2. Joseph Needham, *La science chinoise et l'Occident* (Paris: Seuil, 1973), 15.
3. Basarab Nicolescu, *La science, le sens et l'évolution: Essai sur Jakob Boehme* (Paris: Félin, 1988), trans. Rob Baker as *Science, Meaning, & Evolution: The Cosmology of Jacob Boehme* (New York: Parabola Books, 1991), foreword by Joscelyn Godwin and afterword by Antoine Faivre.
4. Alexandre Kojève, "L'origine chrétienne de la science moderne," in *L'aventure de l'esprit*, vol. 2 of *Mélanges Alexandre Koyré*, ed. Fernand Braudel (Paris: Hermann, 1964), 295–306, foreword by Fernand Braudel.
5. Ibid., 297.
6. Ibid., 298.
7. Ibid., 301.
8. Ibid., 302.
9. Ibid., 303.
10. Ibid., 304–305.
11. Alexandre Leupin, *Fiction et Incarnation: Littérature et théologie au Moyen Age* (Paris: Flammarion, 1993).
12. Ibid., 43. The original French term for "anti-writing" is *antécrit*, a neologism invented by the author, whose significance is that of Tertulian's expression "pre-scriptio" (what was written before).
13. Steven Louis Goldman, "Alexander Kojève on the Origin of Modern Science: Sociological Modelling Gone Awry," *Studies in History and Philosophy of Science* 6 no. 2 (1975): 113–24.
14. Galilei, *Dialogue*, 129.
15. Michel Camus et al., "Levels of Representation and Levels of Reality: Towards an Ontology of Science," in *The Concept of Nature in Science and Theology*, ed. Niels H. Gregersen, Michael W. S. Parsons, and Christoph Wassermann (Geneva: Labor et Fides, 1998), 2:94–103.
16. Horia Bădescu and Basarab Nicolescu, eds., *Stéphane Lupasco: L'homme et l'oeuvre*, (Monaco: Rocher, 1999).
17. Thierry Magnin, *Entre science et religion: Quête de sens dans le monde présent* (Monaco: Rocher, 1998), foreword by Basarab Nicolescu and afterword by Henri Manteau-Bonamy.

18. Nicolescu, *Science, le sens et l'évolution*, 39–53.

19. Anthony Crafton, *Cardano's Cosmos: The Worlds and Works of a Renaissance Astrologer* (Cambridge, MA: Harvard University Press, 1999).

20. Gérard Simon, *Kepler astronome astrologue* (Paris: Gallimard, 1979).

21. B. J. T. Dobbs, *The Janus Faces of Genius: The Role of Alchemy in Newton's Thought* (Cambridge: Cambridge University Press, 1991).

22. C. P. Snow, *The Two Cultures and the Scientific Revolution* (New York: Cambridge University Press, 1961).

23. Basarab Nicolescu, *La transdisciplinarité, manifeste* (Monaco: Rocher, 1996), trans. Karen-Claire Voss as *Manifesto of Transdisciplinarity* (New York: SUNY Press, 2002).

24. Jacques Hadamard, *An Essay on the Psychology of Invention in the Mathematical Field* (New York: Princeton University Press, 1945).

25. Henry Corbin, *Face de Dieu, face de l'homme: Herméneutique et soufisme* (Paris: Flammarion, 1983).

26. Suzi Gablik, *Has Modernism Failed?* (New York: Thames & Hudson, 2004).

27. Michael Kustow, *Peter Brook: A Biography* (New York: St. Martin's Press, 2005).

28. Michel Camus, "The Hidden Hand between Poetry and Science," in *Transdisciplinarity: Theory and Practice*, ed. Basarab Nicolescu (Creskill, NJ: Hampton Press, 2008).

29. Mircea Eliade, *Images and Symbols: Studies in Religious Symbolism*, trans. Philip Mairet (Princeton, NJ: Princeton University Press, 1991), 170.

30. Christian Moraru, *Cosmodernism: American Narrative, Late Globalization, and the New Cultural Imaginary* (Ann Arbor: The University of Michigan Press, 2011).

31. Mircea Eliade, *L'épreuve du labyrinthe: Entretiens avec Claude-Henri Roquet* (Paris: Pierre Belfond,1978), 175–76.

32. Bryan S. Rennie, ed., *Changing Religious Worlds: The Meaning and End of Mircea Eliade* (Albany, NY: SUNY Press, 2001).

33. Michel Cazenave, "La place et la fonction du transculturel dans le monde contemporain," *Transdisciplinary Studies: Science, Spirituality, Society* 1 (2011): 173.

34. C. Cohen, *La femme des origines: Images de la femme dans la préhistoire occidentale* (Paris: Belin-Herscher, 2003).

35. George Steiner, "Penser Europe," in *L'Europe en quête d'harmonie*, ed. Aude Fonquernie (Mazille, France: La Maison sur le Monde, 2001), 66.

36. Adonis, *La prière et l'épée: Essais sur la culture arabe*, trans. Leïla Khatib and Anne Wade Minkowski (Paris: Mercure de France, 1993), 143–46.

37. Raimon Panikkar, *The Intrareligious Dialogue* (Mahwah, NJ: Paulist Press, 1999).

CHAPTER TWO

1. Djalâl-ud-Dîn Rûmî, *Le livre du dedans (Fîhi-mâ-fîhi)*, trans. Eva de Vitray-Meyerovitch (Paris: Sindbad, 1975).

2. Albert Einstein, "Remarques préliminaires sur les concepts fondamentaux," in *Louis de Broglie: physicien et penseur* (Paris: Albin Michel, 1953), 7.

3. Jacob Boehme, "De la Base sublime et profonde des six points théosophiques," trans. Louis-Claude de Saint-Martin, in *Jacob Boehme: Cahiers de l'Hermetisme*, ed. Antoine Faivre and Frédérick Tristan (Paris: Albin Michel, 1977), 134.

4. Farid ud-Din Attar, *The Conference of the Birds*, trans. C. S. Nott (Boulder, CO: Shambhala, 1971).

5. Basarab Nicolescu, "Science as Testimony," in *Science and The Boundaries of Knowledge: The Prologue of Our Cultural Past*, Final Report REX-86/WS-23 (Paris: UNESCO, 1986), 9–29.

6. Philippe Frank, *Einstein: sa vie et son temps* (Paris: Albin Michel, 1950), 9.

7. Friedrich Dürrenmatt, *Albert Einstein*, trans. Françoise Stokke (Lausanne: Éditions de l'Aire, 1982), 32.

8. Jacob Boehme, *L'Aurore naissante*, trans. Le Philosophe Inconnu (Milano: Arché, 1977), 60.

9. P. D. Ouspensky, *Fragments d'un enseignement inconnu*, trans. Philippe Lavastine (Paris: Stock, 1978), 396.

10. Boehme, *L'Aurore naissante*, 313.

11. Ibid., 396.

12. Ibid., 336.

13. Saint Jean de la Croix, "La Montée du Carmel," in *Oeuvres spirituelles* (Paris: Seuil, 1954); see also Jean Baruzi, *Saint Jean de la Croix et le problème de l'expérience mystique* (Paris: Librairie Félix Alcan, 1931).

14. Saint Jean de la Croix, "La Montée du Carmel," 34.

15. Albert Einstein, *Comment je vois le Monde* (Paris: Flammarion, 1934), 35.

16. Ludovic de Gaigneron, *L'Image ou le drame de la nullité cosmique* (Paris: Le Cercle du Livre, 1956).

17. Basarab Nicolescu, *Nous, la particule et le monde*, 3rd ed. (Brussels: E. M. E. & InterCommunications, 2012).

18. Albert Einstein to Max Born, letter, December 4, 1926, quoted in Gerald Holton, *L'Imagination scientifique* (Paris: Gallimard, 1981), 81.

19. Nicolescu, *Science, Meaning, & Evolution*.

20. Boehme, *L'Aurore naissante*, 63.

21. Boehme, "De la Base," 122.

22. Boehme, *L'Aurore naissante*, 178.

23. Ibid., 188.

24. Ibid., 392.

25. Ibid., 124.

26. Boehme, "De Electione Gratiae," in *Jacob Boehme: Cahiers de l'Hermetisme*, ed. Antoine Faivre and Frédérick Tristan (Paris: Albin Michel, 1977), 68.

27. Boehme, *L'Aurore naissante*, 58.

28. Ibid., 319.

29. Ibid., 315.

30. Stéphane Lupasco, *Les trois matières* (Paris: Julliard 10:18, 1970).

31. Boehme, "De la Base," 133.

32. Jacob Boehme, *Mysterium Magnum*, 4 vols., trans. N. Berdiaev (Paris: Éditions d'Aujourd'hui, 1978).

33. N. Berdiaeff, "L'"Ungrund' et la liberté," in Boehme, *Mysterium Magnum*, trans. N. Berdiaev (Paris: Éditions d'Aujourd'hui, 1978), 1:26.

34. Boehme, *L'Aurore naissante*, 429.

35. See, for example, René Guénon, *La Grande Triade* (Paris: Gallimard, 1974), and *Le Symbolisme de la Croix* (Paris: Vega, 1970).

36. Gilbert Durand, *L'Imagination symbolique* (Paris: Quadrige/PUF, 1984), 13.

37. Ouspensky, *Fragments d'un enseignement inconnu*, 400–401.

38. Alfred Korzybski, *Science and Sanity* (Lakeville, CT: The International Non-Aristotelian Library Publishing Company, 1958).

39. Gerald Holton, *L'imagination scientifique*, trans. Jean-François Roberts (Paris: Gallimard, 1981), originally published as *The Scientific Imagination: Case Studies* (Cambridge: Cambridge University Press, 1978); see also Gerald Holton, *Thematic Origin of Scientific Thought: From Kepler to Einstein* (Cambridge, MA: Harvard University Press, 1973).

40. Gerald Holton, "Les thêmata dans la pensée scientifique," in *L'Imagination scientifique*, 30.

41. Angèle Kremer-Marietti, "Gerald Holton," in *Dictionnaire des philosophes*, ed. Denis Huisman (Paris: PUF, 1984), 1:1251.

42. Holton, "Les thêmata dans la pensée scientifique," 30.

43. Jakob Boehme, "Mysterium pansophicum," trans. Louis-Claude de Saint-Martin, in *Jacob Boehme: Cahiers de l'Hermétisme*, ed. Antoine Faivre and Frédérick Tristan (Paris: Albin Michel, 1977), 190.

44. Boehme, *L'Aurore naissante*, 72.

45. Geoffrey F. Chew, "Bootstrap: A Scientific Idea?" *Science* 161 (1968): 762–65.

46. René Guénon, *Le Règne de la quantité et les signes des temps* (Paris: Gallimard, 1970), 10.

47. *Science and the Boundaries of Knowledge: The Prologue of Our Cultural Past*, Final Report REX-86/WS-2 (Paris: UNESCO, 1986).

48. "Venice Declaration: Final Communiqué of the Symposium Science and the Boundaries of Knowledge: The Prologue of Our Cultural Past," CIRET: Centre International de Recherches et Études Transdisciplinaires, March 7, 1986, last modified June 1994, accessed March 25, 2013, http://ciret-transdisciplinarity.org/bulletin/b2c4_en.php.

CHAPTER THREE

1. Galileo Galilei, *Dialogue Concerning the Two Chief World Systems, Ptolemaic and Copernican*, trans. Stillman Drake (Berkeley: University of California Press, 1962). Foreword by Albert Einstein.

2. Basarab Nicolescu, "La mécanique quantique, l'évolution et le progrès," in *Sciences et conscience*, ed. Alain Houziaux (Paris: Albin Michel, 1999), 23–30.

3. Vincent Descombes, *Le complément du sujet* (Paris: Gallimard, 2004).

4. Max Planck, *Autobiographie scientifique*, trans. André George (Paris: Albin Michel, 1960).

5. Nicolescu, *Nous, la particule*.

CHAPTER FOUR

1. Max Planck, *Initiations à la physique* (Paris: Flammarion, 1941), 177–78.
2. Attar, *Conference of the Birds*.
3. Nicolescu, *Science, Meaning, & Evolution*.
4. Planck, *Autobiographie scientifique*, 73.
5. Ibid., 76.
6. Lilian Silburn, *Instant et cause: Le discontinu dans la pensée philosophique de l'Inde* (Paris: Vrin, 1995).
7. Max Jammer, *The Conceptual Development of Quantum Mechanics* (New York: McGraw-Hill Book Company, 1966), 324.
8. C. F. von Weizsäcker, "A Reconstruction of Quantum Theory," in *Quantum Theory and the Structure of Time and Space*, ed. L. Castell and M. Drieschner (Munich: Carl Hanser Verlag, 1979), 3:25.
9. Emilio Segré, *Les Physiciens modernes et leurs découvertes* (Paris: Fayard, 1984), 88.
10. Planck, *Autobiographie scientifique*, 94.
11. Ibid., 11.
12. D. Finkelstein, "Holistic Methods in Quantum Logic," in *Quantum Theory and the Structure of Time and Space*, ed. L. Castell and M. Drieschner (Munich: Carl Hanser Verlag, 1979), 3:41.
13. Bernard d'Espagnat, "Nonseparability and the Tentative Descriptions of Reality," *Physics Reports* 110, no. 4 (1987): 203.
14. J. S. Bell, *Speakable and Unspeakable in Quantum Mechanics: Collected Papers on Quantum Philosophy* (Cambridge: Cambridge University Press, 1987).
15. Bernard d'Espagnat, *À la Recherche du Réel* (Paris: Gauthier-Villars, 1981); see also Bernard d'Espagnat, "The Quantum Theory and Reality," *Scientific American* 241, no. 5 (1979): 128–40.
16. Alain Aspect, Philippe Grangier, and Gérard Roger, Experimental Tests of Realistic Local Theories via Bell's Theorem, *Physical Review Letters* 47, no. 7 (1981): 460; Alain Aspect, Jean Dalibard, and Gérard Roger, Experimental Test of Bell's Inequalities Using Time-Varying Analyzers, *Physical Review Letters* 49, no. 25 (1982): 1804.
17. d'Espagnat, *À la Recherche du Réel*, 26.
18. Max Tegmark and John Archibald Wheeler, "100 Years of the Quantum," *Scientific American*, 2 (2001): 68.
19. Heinz R. Pagels, *The Cosmic Code* (Toronto: Bantam Books, 1983), 126.
20. Henry P. Stapp, *Mind, Matter and Quantum Mechanics* (New York: Springer Verlag, 1993); Roger Penrose, *Shadows of the Mind: A Search for the Missing Science of Consciousness* (Oxford: Oxford University Press, 1995); John Eccles, *How the Self Controls Its Brain* (New York: Springer Verlag, 1994).

CHAPTER FIVE

1. Abdus Salam, interview by Michel Random, "L'Unité des quatre énergies de l'univers," *3e Millénaire* 2 (1982): 16.

2. Steven Weinberg, "Unified Theories of Elementary-Particle Interaction," *Scientific American* 231, no. 1 (1974): 50–59.

3. Abdus Salam, cited in Pagels, *The Cosmic Code*, 266–67.

4. Leon Lederman, *The God Particle: If the Universe Is the Answer, What Is the Question?* (Boston: Houghton Mifflin Company, 1993), 348.

5. D. Z. Freedman and P. van Nieuwenhuizen, "La Supergravité et l'unification des lois de la physique," in *Les Particules élémentaires* (Paris: Bibliothèque "Pour la science," 1983), 193–206.

6. Paul Davies, *Superforce: The Search for a Great Unified Theory of Nature* (New York: Simon & Schuster, 1984), 150–68.

7. Abdus Salam, "The Nature of the 'Ultimate' Explanation in Physics," in *Scientific Explanation: Papers Based on Herbert Spencer Lectures Given in the University of Oxford*, ed. A. F. Heath (Gloucestershire: Clarendon, 1981).

8. Lederman, *The God Particle*, 348.

9. Basarab Nicolescu, "Relativité et physique quantique," in *Dictionnaire de l'ignorance*, ed. Michel Cazenave (Paris: Albin Michel, 1998), 108–20.

10. G. D. Couglan and J. E. Dodd, *The Ideas of Particle Physics* (Cambridge: Cambridge University Press, 1991), 214–18.

11. Sheldon L. Glashow, *Le Charme de la physique*, trans. Olivier Colardelle (Paris: Albin Michel, 1997), 14.

12. Murray Gell-Mann, quoted in Couglan and Dodd, *Ideas of Particle Physics*, 214–18.

13. Laurent Nottale, *Fractal Space-Time and Microphysics: Towards a Theory of Scale Relativity* (Singapore: World Scientific, 1992).

14. Murray Gell-Mann, *Le Quark et le jaguar*, trans. Gilles Minot (Paris: Flammarion, 1997), 150–51.

15. An excellent introduction to superstring theory and its history is Brian Greene, *L'Univers élégant: Une révolution scientifique: de l'infiniment grand à l'infiniment petit, l'unification de toutes les théories de la physique*, trans. Céline Laroche (Paris: Robert Laffont, 2000).

16. Greene, *L'Univers élégant*, 276.

17. Glashow, *Le Charme de la physique*, 300–301.

18. Edward Witten, "Duality, Spacetime and Quantum Mechanics," *Physics Today* 50, no. 5 (May 1997): 28–33.

19. Edwin A. Abbott, *Flatland: Une aventure à plusieurs dimensions*, trans. Elisabeth Gille (Paris: Denoël, 1984).

20. Alain Connes, André Lichnerowicz, and Marcel Paul Schützenberger, *Triangle de pensées* (Paris: Odile Jacob, 2000). See chapter 8, "Réflexions sur le temps."

21. Leonard Susskind, *The Cosmic Landscape: String Theory and the Illusion of Intelligent Design* (New York: Bay Books, 2006).

22. Ibid., 111–130.

23. Lee Smolin, *The Trouble with Physics: The Rise of String Theory, the Fall of a Science, and What Comes Next* (London: Penguin Books, 2006), 121.

24. Ibid., 18–19.

25. Lederman, *The God Particle*, 193–98.

26. Smolin, *The Trouble with Physics*, xvi.

27. John D. Barrow, *La Grande Théorie: Les limites d'une explication globale en physique*, trans. Michel Cassé and Guy Paulus (Paris: Albin Michel, 1994).
28. Ernest Nagel and James R. Newman, *Gödel's Proof* (New York: New York University Press, 1958).
29. Basarab Nicolescu, "Gödelian Aspects of Nature and Knowledge," in *Systems: New Paradigms for the Human Sciences*, ed. Gabriel Altmann and Walter A. Koch (Berlin: De Gruyter, 1998).
30. Greene, *L'Univers élégant*, 418, 420.

CHAPTER SIX

1. Michio Kaku, *Hyperspace: A Scientific Odyssey through Parallel Universes, Time Warps, and the Tenth Dimension* (Oxford: Oxford University Press, 1994).
2. Abbott, *Flatland*.
3. Michael S. Morris, Kip S. Thorne, and Ulvi Yurtsever, "Wormholes, Time Machines, and the Weak Energy Condition," *Physical Review Letters* 61, no. 13 (1988): 1446–49.
4. Rudy Rucker, *La quatrième dimension*, trans. Christian Jeanmougin (Paris: Seuil, 1985), foreword by Martin Gardner, 81–90.
5. Charles Howard Hinton, *The Fourth Dimension* (Salem, NH: Ayer Company, 1986, facsimile of the 1912 edition, London: George Allen & Unwin Ltd). First edition published in 1904.
6. Charles Howard Hinton, *Scientific Romances*, 2nd series (Leeds: Celephaïs Press, 2008, first edition London: Swan Sonnenschein, 1896), 33, available at http://fr.scribd.com/doc/8552379/Hinton-Scientific-Romances-2.
7. Ibid.
8. P. D. Ouspensky, *Tertium Organum: A Key to the Enigmas of the World*, trans. Nicholas Bessaraboff and Claude Bragdon (Rochester, NY: Manas Press, 1920, first edition Saint Petersburg: N. P. Taberio, 1911).
9. J. B. Priestley, *Man and Time* (London: Aldus Books, 1964).
10. Jean Clair, "Ouspensky et l'espace néo-platonicien," in *Malevitch*, ed. Jean-Claude Marcadé (Lausanne: L'Age d'Homme, 1979), 15–29.
11. Linda Dalrymple Henderson, *The Fourth Dimension and Non-Euclidean Geometry in Modern Art* (Princeton, NJ: Princeton University Press, 1983); "The Merging of Time and Space: The Fourth Dimension in Russia from Ouspensky to Malevitch," *The Structurist* 15/16 (1975/1976): 97–108.
12. Hermann Minkowski, "Raum und Zeit," paper presented at the Eightieth Meeting of German Physicists and Physicians, Köln, September 21, 1908, *Physikalische Zeitschrift* 10 (1909).

CHAPTER SEVEN

1. A. S. Eddington, *The Philosophy of Physical Science* (1939; repr., Cambridge: Cambridge University Press, 1949), 57.
2. E. Whittaker, *Eddington's Principle in the Philosophy of Science* (Cambridge: Cambridge University Press, 1951).

3. A. S. Eddington, *La Nature du monde physique* (Paris: Payot, 1929), 286.

4. G. F. Chew, "Theory of Strong Coupling of Ordinary Particles," in *Proceedings of the Ninth International Annual Conference on High Energy Physics* (Kiev, 1959), 332; Maurice Jacob and Geoffrey F. Chew, *Strong-Interaction Physics: A Lecture Note Volume* (New York: W. A. Benjamin, 1964).

5. G. F. Chew and S. Mandelstam, "Theory of the Low-Energy Pion-Pion Interaction: II," *Nuovo Cimento* 19 (1961): 752–76.

6. James T. Cushing, *Theory Construction and Selection in Modern Physics: The S Matrix* (Cambridge: Cambridge University Press, 1990).

7. G. F. Chew, "Impasse for the Elementary-Particle Concept," in *The Great Ideas Today 1974* (London: Encyclopaedia Britannica, 1974), 114–15.

8. Ibid., 119; see also G. F. Chew, "Hadron 'Bootstrap': Triumph or Frustration?" *Physics Today* 23, no. 10 (1970): 23–28.

9. Chew, "'Bootstrap.'"

10. G. F. Chew, *S-Matrix Theory of Strong Interaction* (New York: W. A. Benjamin, 1961).

11. Salam, "The Nature of the 'Ultimate.'"

12. R. H. Dicke, «Dirac's Cosmology and Mach's Principle,» *Nature* 192 (1961): 440–41.

13. Dicke, «Dirac's Cosmology»; P. A. M. Dirac, «The Cosmological Constants,» *Nature* 139 (1937): 351–52.

14. John Barrow, Frank Tipler, and Marie-Odile Monchicourt, *L'Homme et le cosmos: Le Principe anthropique en astrophysique moderne* (Paris: Imago, 1984), 79–80, afterword by Hubert Reeves.

15. Stephen Hawking, "Les Objectifs de la physique théorique," *CERN Courier* 21 (January-February 1981): 4.

16. Z. Maki, Y. Ohnuki, and S. Sakata, "Remarks on a New Concept of Elementary Particles and the Method of the Composite Model," *Proceedings of the International Conference on Elementary Particles*, ed. Y. Tanikawa (Kyoto: Kyoto University, 1965), 109–15.

17. Y. Ne'eman, "Concrete versus Abstract Theoretical Models," in *Interaction between Science and Philosophy*, ed. Y. Elkana (Atlantic Highlands, NJ: Humanities Press, 1975), 16.

18. Nigel Calder, *The Key of the Universe* (London: British Broadcasting Corporation, 1977), 93.

19. James T. Cushing, "Is There Just One Possible World? Contingency vs. the Bootstrap," *Studies in the History and Philosophy of Science* 16 (1985): 31–48.

20. George Gale, "Chew's Monadology," *History of Ideas* 35 (1974): 339–48; "Leibniz, Chew and Wheeler on the Identity of Physical and Philosophical Inquiry," *Review of Metaphysics* 29 (1975): 323–33.

21. Jean-Paul Dumont, "Anaxagore," in *Dictionnaire des philosophes*, ed. Denis Huisman (Paris: PUF, 1984), 1:77–81.

22. Julia H. Graisser and T. K. Graisser, "Partons in Antiquity," *American Journal of Physics* 45, no. 5 (1977).

23. Fritjof Capra, *The Tao of Physics: An Exploration of the Parallels between Modern Physics and Eastern Mysticism* (Berkeley: Shambala, 1975).

24. J. P. Merlo, "Physique et philosophie," preprint, submitted 1975, Centre d'Études Nucléaires de Saclay.

CHAPTER EIGHT

1. For an excellent introduction to complexity philosophy, see Paul Cilliers, *Complexity and Postmodernism: Understanding Complex Systems* (London: Routledge, 1998).
2. Ervin Laszlo, *Le Systémisme: Vision nouvelle du monde* (Paris: Pergamon Press, 1981).
3. Edgar Morin, *La Méthode*, Vol. 1, *La Nature de la Nature* (Paris: Seuil, 1977), 149.
4. Ludwig von Bertalanffy, *Théorie générale des systèmes* (Paris: Dunod, 1973).
5. Murray Gell-Mann, *Le Quark et le jaguar*, trans. Gilles Minot (Paris: Flammarion, 1997).
6. Laszlo, *Le Systémisme*, 59.
7. Erich Jantsch, *The Self-Organizing Universe* (Oxford: Pergamon Press, 1980).
8. Basarab Nicolescu, "Quelques réflexions sur la pensée atomiste et la pensée systémique," 3e *Millénaire* 7 (1983): 20–26.
9. Edgar Morin, *La Méthode*, Vol. 3, *La Connaissance de la connaissance, livre 1* (Paris: Seuil, 1986).
10. Basarab Nicolescu, *La transdisciplinarité: Manifeste* (Monaco: Rocher, 1996).
11. Werner Heisenberg, *Philosophie: Le manuscrit de 1942*, trans. and with an introduction by Catherine Chevalley (Paris: Seuil, 1998). A translation in English of this book can be found at http://werner-heisenberg.unh.edu/t-OdW-english.htm#seg01.
12. Steven Weinberg, *Les Trois Premières Minutes de l'univers* (Paris: Seuil, 1978), 173.
13. A. H. Guth, "Inflationary Universe: A Possible Solution to the Horizon and Flatness Problems," *The Physical Review* D23, no. 2 (1981): 347–56.
14. Paul Davies, *Superforce*, 195.
15. Weinberg, *Les Trois Premières Minutes*, 179.
16. Pierre Teilhard de Chardin, *Oeuvres*, 10 vols. (Paris: Seuil, 1965–1970).
17. Frank J. Tipler, *The Physics of Immortality: Modern Cosmology, God and the Resurrection of the Dead* (New York: Doubleday, 1994).

CHAPTER NINE

1. Solomon Marcus, *Introduction mathématique à la linguistique structurale* (Paris: Dunod, 1967).
2. Charles Sanders Peirce, *Values in a Universe of Chance: Selected Writings of Charles S. Peirce*, ed. Philip P. Wiener (New York: Dover Publications, 1966), 133.
3. Ibid.
4. Don D. Roberts, *The Existential Graphs of Charles S. Peirce* (The Hague: Mouton Humanities Press, 1973).
5. Charles S. Peirce, *Écrits sur le signe*, ed. Gérard Deledalle (Paris: Seuil, 1978), 22.

6. Roberts, *The Existential Graphs*, 115.
7. Peirce, *Écrits sur le signe*, 69.
8. Peirce, *Values*, 359.
9. Peirce, *Écrits sur le signe*, 22.
10. Ibid., 72.
11. Ibid., 23–24.
12. Ibid., 70.
13. Ibid., 148.
14. Roberts, *The Existential Graphs*, 147.
15. Ibid.
16. Peirce, *Values*, 147.
17. Peirce, *Écrits sur le signe*, 69–70.
18. Ibid., 24.
19. Ibid., 98–99.
20. Ibid., 115.
21. Ibid., 74.
22. Deledalle in Peirce, *Écrits sur le signe*, 209.
23. Isabel Stearns, "Firstness, Secondness, and Thirdness," in *Studies in the Philosophy of Charles Sanders Peirce*, ed. Philip P. Wiener and Frederic H. Young (Cambridge, MA: Harvard University Press, 1952), 198.
24. Peirce, *Values*, 93.
25. Ibid., 107.
26. Ibid., 148.
27. Stéphane Lupasco, *Les trois matières* (Paris: Julliard, 1970), 128.
28. Peirce, *Écrits sur le signe*, 30.
29. Ibid., 138–39.
30. Stearns, "Firstness, Secondness, and Thirdness," 196.
31. Thirring, "Do the Laws of Nature Evolve?" 131–36.
32. W. L. Rosensohn, *The Phenomenology of Charles S. Peirce: From the Doctrine of Categories to Phaneroscopy* (Amsterdam: B. R. Grüner, 1974), 69.
33. Peirce, *Écrits sur le signe*, 249.
34. Ibid., 252.
35. Ibid., 61.
36. Ibid., 62.
37. Peirce, *Values*, 119.
38. Jean-François Malherbe, *Le Nomade polyglotte* (Saint-Laurent, QC: Bellarmin, 2000).

CHAPTER TEN

1. Niels Bohr, *Essays 1958–1962 on Atomic Physics and Human Knowledge* (New York: Interscience Publishers, 1963).
2. Alfred Korzybski, *Science and Sanity* (1933; Lakeville, CN: The International Non-Aristotelian Library Publishing Company, 1958).
3. Bădescu and Nicolescu, *Stéphane Lupasco*; see also Basarab Nicolescu, "Stéphane Lupasco (1900–1988)," in *Universalia 1989* (Paris: Encyclopaedia Universalis,

1989), section "Vies et portraits," and "Stéphane Lupasco (1900–1988)," in *Les oeuvres philosophiques: Dictionnaire* (Paris: Encyclopédie Philosophique Universelle [PUF], 1992).

4. Lupasco, *Les trois matières*, 58.

5. Stéphane Lupasco, *Du devenir logique et de l'affectivité*: Vol. 1, *Le dualisme antagoniste et les exigences historiques de l'esprit*, and Vol. 2, *Essai d'une nouvelle théorie de la connaissance: La physique macroscopique et sa portée philosophique*, (Paris: Vrin, 1935). Volume 2 was Lupasco's PhD thesis, complementary to Volume 1. A second edition of both volumes was published in 1973.

6. Stéphane Lupasco, *L'expérience microphysique et la pensée humaine*, 2nd ed. (Monaco: Rocher, 1989), preface by Basarab Nicolescu.

7. Ibid., 20, 7, 14.

8. Stéphane Lupasco, *Le principe d'antagonisme et la logique de l'énergie: Prolégomènes à une science de la contradiction* (Paris: Hermann, 1951).

9. T. A. Brody, "On Quantum Logic," *Foundations of Physics* 14, no. 5 (1984): 409–30.

10. Stéphane Lupasco, *Psychisme et sociologie* (Paris: Casterman, 1978), 10.

11. Vintila Horia, *Viaje a los centres de la Tierra: Encuesta sobre el Estado Actual del Pensamiento, Las Artes y Las Ciencias* (Barcelona: Ediciones de Nuevo Arte Thor, 1987), 298, 305, 306.

12. Gilbert Durand, *L'imagination symbolique* (Paris: PUF, 1984), 95–96; see also Gilbert Durand, "L'Anthropologie et les structures du complexe," in *Stéphane Lupasco: L'homme et l'oeuvre*, ed. Horia Bădescu and Basarab Nicolescu (Monaco: Rocher, 1999), 61–74.

13. Lupasco, *Les trois matières*, 75.

14. Ibid., 52.

15. Lupasco, *Le principe d'antagonisme*, 70.

16. R. Penrose and M. A. H. McCollum, "Twistor Theory: An Approach to the Quantization of Fields and Space-Time," *Physics Reports* 6C, no. 4 (1973): 243.

17. Lupasco, *Le principe d'antagonisme*, 114.

18. Ibid., 105

19. Costin Cazaban, "Temps musical/espace musical comme fonctions logiques," in *L'esprit de la musique: Essais d'esthétique et de philosophie*, ed. Hugues Dufourt, Joël-Marie Fouquet, and François Hurard (Paris: Klincksieck, 1992).

20. Pompiliu Craciunescu, *Vintila Horia: Translittérature et Réalité* (Veauche, France: L'Homme Indivis, 2008).

21. Adrian Cioroianu, "Stéphane Lupasco temps et contradiction: Vers une nouvelle logique de l'histoire?" in *À la confluence de deux cultures: Lupasco aujourd'hui*, ed. Basarab Nicolescu (Paris: Oxus, 2010), 20–39.

22. Jean-François Malherbe, "Esquisse d'une histoire de l'éthique à l'aune du tiers inclus," in *À la confluence de deux cultures: Lupasco aujourd'hui*, ed. Basarab Nicolescu (Paris: Oxus, 2010), 40–53.

23. Edgar Morin, "Lupasco et les pensées qui affrontent les contradictions," in *À la confluence de deux cultures: Lupasco aujourd'hui*, ed. Basarab Nicolescu (Paris: Oxus, 2010), 103–29.

24. Michel Cazenave, "Psychologie et tiers inclus," in *À la confluence de deux cultures: Lupasco aujourd'hui*, ed. Basarab Nicolescu (Paris: Oxus, 2010),

130–34; Jean-Louis Revardel, "Lupasco et la translogique de l'affectivité," in À la confluence de deux cultures: Lupasco aujourd'hui, ed. Basarab Nicolescu (Paris: Oxus, 2010), 135–58.

25. Paul Ghils, "Langage pur, langage impur: Du mythe de l'origine à la pragmatique de la contradiction," in À la confluence de deux cultures: Lupasco aujourd'hui, ed. Basarab Nicolescu (Paris: Oxus, 2010), 159–96.

26. Michel De Caso, "La mise à jour des Lumières: Tiers inclus, niveaux de Réalité et Rectoversion," in À la confluence de deux cultures: Lupasco aujourd'hui, ed. Basarab Nicolescu (Paris: Oxus, 2010), 228–49.

27. À la confluence de deux cultures: Lupasco aujourd'hui, Colloque International UNESCO, Paris, March 24, 2010.

28. See, for example: Camus et al., "Levels of Representation"; Magnin, Entre science et religion; Bernard Morel, Dialectiques du Mystère (Paris: La Colombe, 1962); Ioan Chirila, "Ternaire et Trinité, homogène, hétérogène et l"état T': Une évaluation théologique du discours lupascien sur le tiers inclus," in À la confluence de deux cultures: Lupasco aujourd'hui, ed. Basarab Nicolescu (Paris: Oxus, 2010), 54–78; Thierry Magnin, "L'unité des antagonismes dans l'histoire de la théologie catholique," in À la confluence de deux cultures: Lupasco aujourd'hui, ed. Basarab Nicolescu (Paris: Oxus, 2010), 79–102; Doru Costache, "Logos, Evolution, and Finality in Anthropological Research: Towards a Transdisciplinary Solution," in Science and Religion: Antagonism or Complementarity? ed. Basarab Nicolescu and Magda Stavinschi (Bucharest: XXI Eonul dogmatic, 2003), 241–60.

29. Jean-Jacques Wunenburger, La raison contradictoire: Sciences et philosophie modernes: La pensée du complexe (Paris: Albin Michel, 1990).

30. Gilles Gaston Granger, L'irrationnel (Paris: Odile Jacob, 1998).

31. David Gross, "On the Uniqueness of Field Theories," in A Passion for Physics: Proceedings of the G. F. Chew Jubilee, September 29, 1984, ed. Carleton DeTar, J. Finkelstein, and Chung-I Tan (Singapore: World Scientific, 1985), 128–36.

32. Edgar Morin, La Méthode, Vol. 1; Vol. 2, La Vie de la Vie (Paris: Seuil, 1980); Vol. 3.

33. Edgar Morin, Science avec conscience (Paris: Fayard, 1982), 118.

34. Planck, Initiations à la physique, 6.

35. Antonio R. Damasio, Looking for Spinoza: Joy, Sorrow, and the Feeling Brain (San Diego, CA: Harcourt, 2003).

36. Jean-François Malherbe, "Jeux de langage" et "tiers inclus": De nouveaux outils pour l'éthique appliquée (Sherbrooke, QC: Université de Sherbrooke, GGC Éditions, 2000).

CHAPTER ELEVEN

1. Umberto Eco, Les limites de l'interprétation (Paris: Grasset, 1992), 58.
2. Ibid., 61.
3. Ibid., 370.
4. Wolfgang Pauli to Markus Fierz, letter, August 12, 1948, in Wolfgang Pauli: Wissenschaftlicher Briefwechsel, Band IV, Teil I: 1940–1949, ed. K. von Meyenn (Berlin: Springer, 1993), 559.

5. Wolfgang Pauli to Markus Fierz, letter, November 3, 1948, quoted in K. V. Laurikainen, *Beyond the Atom: The Philosophical Thought of Wolfgang Pauli* (Berlin: Springer, 1988), 74.

6. Wolfgang Pauli, to Markus Fierz, letter, June 3, 1952, quoted in Laurikainen, *Beyond the Atom*, 141.

7. Christine Maillard, *Les sept sermons aux morts de Carl Gustav Jung* (Nancy: Presses Universitaires de Nancy, 1993). Jung's text is quoted in its entirety at the beginning of this book.

8. C. G. Jung, *Ma vie: Souvenirs, rêves et pensées*, ed. Aniela Jaffé (Paris: Gallimard, 1992), 232.

9. Wolfgang Pauli to Abraham Pais, letter, August 17, 1950, In K. von Meyenn, *Wolfgang Pauli*, Vol. 1, *1950–1952* (Berlin: Springer, 1993), 152.

10. Harald Atmanspacher and Hans Primas, "Pauli's Ideas on Mind and Matter in the Context of Contemporary Science," *Journal of Consciousness Studies* 13, no. 3 (2006): 5–50.

11. See, for instance, Wolfgang Pauli to Markus Fierz, letter, October 13, 1951, quoted in Laurikainen, *Beyond the Atom*, 40.

12. Lupasco, *L'expérience microphysique*.

13. Laurikainen, *Beyond the Atom*, 177.

14. Wolfgang Pauli to Markus Fierz, letter, March 5, 1957, quoted in Laurikainen, *Beyond the Atom*, 84–85.

15. Lupasco, *Les trois matières*.

16. C. G. Jung, *Essais sur la symbolique de l'esprit* (Paris: Albin Michel, 1991), 198, 157.

17. C. G. Jung, *Commentaire sur le Mystère de la Fleur d'Or* (Paris: Albin Michel, 1980), 39.

18. Jung, *Essais sur la symbolique*, 223.

19. Ibid., 160.

20. Laurikainen, *Beyond the Atom*, chapter 9, "Quaternity," 125–39.

21. Marie-Louise von Franz, *Nombre et temps: Psychologie des profondeurs et physique moderne* (Paris: La Fontaine de Pierre, 1983).

22. Eco, *Les limites de l'interprétation*, 55–66.

CHAPTER TWELVE

1. Roberto Juarroz, *Nouvelle poésie verticale* (Paris: Lettres Vives, 1984), 29.

2. Nathalie Sarraute, *Pour un oui ou pour un non* (Paris: Gallimard, 1998), 16.

3. Ibid., 57.

4. Marguerite Jean-Blain, *Eugène Ionesco: Mystique ou mal-croyant?* (Brussels: Lessius, 2005), 63–64.

5. Eugène Ionesco, *Journal en miettes* (Paris: Mercure de France, 1967), 62.

6. Eugène Ionesco, *Victimes du devoir*, in *Théâtre: Tome I* (1954; Paris, Gallimard, 1984), 159–213.

7. Ibid., 200–01, 203–05.

8. Eugène Ionesco, *Théâtre complet* (Pléiade series, Paris: Gallimard, 1990), CI–CII.

9. Emmanuel Jacquart, ed., "Notice," in *Théâtre complet*, Eugène Ionesco (Paris: Gallimard, 1990), 1500–1502.
10. Wylie Sypher, *Loss of the Self in Modern Literature and Art* (New York: Random House, 1962), 87–109.
11. Ibid., 99.
12. Ibid., 97.
13. Ibid., 100.
14. Ibid., 104–05.
15. Gregorio Morales, *El cadáver de Balzac: Una visión cuántica de la literatura y el arte* (Alicante: Epígono,1998).
16. Leonard Shlein, *Art and Physics: Parallel Visions in Space, Time, and Light* (1991; New York: Harper Perennial, 2007).
17. *The World of Quantum Culture*, ed. Manuel J. Caro and John W. Murphy (Westport, CT: Praeger Publishers, 2002).
18. Gregorio Morales, "Overcoming the Limit Syndrome," in *The World of Quantum Culture*, ed. Manuel J. Caro and John W. Murphy (Westport, CT: Praeger Publishers, 2002).
19. Accessed February 15, 2010, http://www.terra.es/personal2/gmv00000/.
20. Gregorio Morales, "Xaverio's Quantum Aesthetics," in *Xaverio: Estética cuantica, petrales 1997–2000* (Granada: Casa de Granda, 2000). All the quotes in English were taken from the site http://www.terra.es/personal2/gmv00000/ (accessed February 15, 2010).
21. Gregorio Morales, *Canto cuántico* (Granada: Dauro, 2003).
22. Allan Riger-Brown, "Gregorio Morales' Quantum Song: A Poetic Voyage into the Enfolded Order," *ELVIRA: Revista de Estudios Filológicos* 10 (2005).
23. Simona Modreanu, "A Different Approach to the 'Theater of the Absurd' with Special Reference to Eugène Ionesco," *Cultura: International Journal of Philosophy and Axiology* 8, no. 1 (2011): 171–86.
24. Pablo Iglesias Simón, "Experiencias en torno al teatro cuántico desde el lado oeste del Golden Gate," *ADE Teatro* 132 (2010): 210–13.
25. Claude Régy, *Espaces perdus* (Besançon: Les Solitaires Intempestifs, 1998); *L'état d'incertitude* (Besançon: Les Solitaires Intempestifs, 2002).
26. Carol Anne Fischer, "Quantum Theatre: A Language for the Voices of Contemporary Theater" (master's thesis, San Jose State University, May 2004), http://www.tvradiofilmtheater.org/MA/Pages/thesesonline/FischerTHESIS.html.

CHAPTER THIRTEEN

1. Basarab Nicolescu, "Peter Brook and Traditional Thought," trans. David Williams, *Contemporary Theatre Review* 7 (1997): 11–23.
2. Peter Brook, quoted in Gérard Montassier, *Le Fait Culturel* (Paris: Fayard, 1980), 122.
3. Matila Ghyka, *Le nombre d'or: Rites et rythmes pythagoriciens dans le development de la civilisation occidentale* (Paris: Gallimard, 1931), preface by Paul Valéry.
4. Peter Brook, personal discussion, June 1984.

5. Peter Brook, *The Empty Space* (Harmondsworth, UK: Penguin Books, 1977), 58.
6. A. G. H. Smith, *Orghast at Persepolis* (London: Eyre Methuen, 1972), 257.
7. John Heilpern, *Conference of the Birds: The Story of Peter Brook in Africa* (Harmondsworth, UK: Penguin, 1979), 103.
8. Brook, *The Empty Space*, 79.
9. Ibid., 13.
10. Peter Brook, *La Conférence des Oiseaux*, program (Paris, C.I.C.T., 1979), 75.
11. Michel Rostain, "Journal des répétitions de *La Tragédie de Carmen*," in *Les Voies de la Création Théâtrale*, Vol. 13, *Peter Brook*, ed. Georges Banu (Paris: Editions de C.N.R.S., 1985).
12. Smith, *Orghast at Persepolis*, 33.
13. Brook, *The Empty Space*, 127.
14. Smith, *Orghast at Persepolis*, 255.
15. Ibid., 123.
16. Brook, *The Empty Space*, 128.
17. John Heilpern, *Conference of the Birds*, 136.
18. Brook, *The Empty Space*, 14–15.
19. Ibid., 124.
20. Peter Brook, *La Cerisaie*, program (Paris: C.I.C.T., 1981), 109.
21. Peter Brook, interview by Ronald Hayman, *The Times*, August 29, 1970.
22. Brook, *The Empty Space*, 40.
23. Ibid., 65.
24. Ibid., 98.
25. Ibid., 96.
26. Ibid., 105.
27. Zeami, *La tradition secrète de No* (Paris: Gallimard, 1960), 77. The most useful of English translations available is *On the Art of the No Drama: The Major Treatises of Zeami*, trans. J. Thomas Rimer and Yamazaki Masakazu (Princeton, NJ: Princeton University Press, 1984).
28. Peter Brook, quoted in Montassier, *Le Fait Culturel*, 115–16.
29. Heilpern, *Conference of the Birds*, 69.
30. Smith, *Orghast at Persepolis*, 250.
31. Brook, *The Empty Space*, 12–13.
32. Ibid., p. 132.
33. Antonio R. Damasio, "Die Insel des Lachens: Erstes Interview mit Antonio Damasio in Iowa City," *Infonautik: Wissen in Bewegung*, http://www.infonautik.de/damasio.htm.
34. Heilpern, *Conference of the Birds*, 50.
35. Peter Brook, quoted in Smith, *Orghast at Persepolis*, 108.
36. Brook, *The Empty Space*, 154.
37. Ibid., 142.
38. Ibid., 57.
39. Ibid., 152.
40. Ibid., 144.

41. Ibid., 25.
42. Ibid., 150.
43. Zeami, *La tradition secrète de No*, 131.
44. Brook, *The Empty Space*, 130.
45. Ibid., 131.
46. Zeami, *La tradition secrète de No*, 131.
47. Brook, *The Empty Space*, 64.
48. Ibid., 29.
49. Georges Banu, "La Conférence des Oiseaux, ou le chemin vers soi-même," in *Les Voies de La Création Théâtrale*, Vol. 10, *Krejka-Brook*, ed. Denis Bablet (Paris: Editions de C.N.R.S., 1985), 285.
50. Brook, *The Empty Space*, 66.
51. Ibid., 45–46.
52. Ed Menta, *The Magic World behind the Curtain: Andrei Serban in the American Theater* (New York: Peter Lang, 1995).

CHAPTER FOURTEEN

1. André Breton, "Les artistes modernes se soucient moins de beauté que de liberté," interview by André Parinaud, *Arts*, Paris, March 7, 1952. in *Entretiens (1913–1952)*, André Breton (Paris: Gallimard, 1952), 296–97.
2. Lupasco, *Du devenir logique*, Vol. 2.
3. André Breton, interview by Jose M. Valverde, *Correo literario*, Madrid, September 1950, in André Breton, *Entretiens (1913–1952)*, (Paris: Gallimard, 1952), 283.
4. Lupasco, *Logique et contradiction*.
5. Mark Polizzotti, *André Breton* (Paris: Gallimard, 1995), 628.
6. André Breton, *Anthologie de l'humour noir* (Paris: Sagittaire, 1950), 345.
7. Lupasco, *Logique et contradiction*, 61, 135, 131, 203.
8. Ibid, 164, 165, 168, 169.
9. André Breton to Stéphane Lupasco, postcard, September 16, 1955, from the private archives of Alde Lupasco-Massot, reproduced here with the authorization of Alde Lupasco-Massot.
10. Stéphane Lupasco, "Réponse au questionnaire d'André Breton," in *L'art magique*, André Breton with G. Legrand (Paris: Club français du livre, 1957), 76–77.
11. Alde Lupasco-Massot archives.
12. Georges Mathieu, *L'abstraction prophétique* (Paris: Gallimard, 1984), 317–32.
13. Paraphrase of Siger de Brabant in Lydia Harambourg, *Georges Mathieu* (Neuchâtel: Ides et Calandes, 2001).
14. Jean-François Malherbe, "Un choix décisif à l'aube de l'éthique: Parménide ou Héraclite," *Transdisciplinarity in Science and Religion* 5 (2009): 11–36.
15. Georges Mathieu, "Mon ami Lupasco," lecture presented at the International Congress "Stéphane Lupasco: L'homme et l'oeuvre," March 13, 1998, Académie des Inscriptions et Belles Lettres, Paris, in *Stéphane Lupasco: L'homme et l'oeuvre*, ed. Horia Bădescu and Basarab Nicolescu (Monaco: Rocher, 1999), 13–28.
16. Stéphane Lupasco, "Quelques aperçus sur la logique dynamique du contradictoire," *United States Lines: Paris Review*, ed. Georges Mathieu (Tours: Arrault

et Cie, 1953). *United States Lines: Paris Review* included contributions by Louis de Broglie, Norbert Wiener, Serge Lifar, A. Rolland de Renéville, and others. Fifteen thousand copies of the magazine were distributed. See also Stéphane Lupasco, "Le principe d'antagonisme et l'art abstrait," *Ring des Arts*, 1960. *Ring des Arts* includes illustrations by Georges Mathieu and contributions by Jean-François Revel, Pierre Restany, Abraham Moles, Georges Mathieu, Alain Bosquet, and others.

17. Georges Mathieu, *Au-delà du Tachisme* (Paris: Julliard, 1963), 164–71.

18. Georges Mathieu, *Mathieu: 50 ans de création* (Paris: Hervas, 2003), 53.

19. Alain Bosquet, *Conversations with Dali* (New York: E. P. Dutton & Co., 1969), 32.

20. Salvador Dali, *Anti-Matter Manifesto*, exhibition booklet (New York: Carstairs Gallery, 1958), quoted in Elliott H. King, "Nuclear mysticism," in *Salvador Dali: Liquid Desire* (Melbourne: National Gallery of Victoria, 2009), 247.

21. Linda Dalrymple Henderson, *The Fourth Dimension and Non-Euclidean Geometry in Modern Art* (Princeton, NJ: Princeton University Press, 1983).

22. Salvador Dali, "Reconstitution du corps glorieux dans le ciel," *Etudes Carmélitaines* 2 (1952): 171–72.

23. Stéphane Lupasco, "Dali and Sub-Realism," in *XXe Siècle*, special issue: "Homage to Salvador Dali," 1980: 117–18.

24. *Les mille et une visions de Dali*, February 19, 1978, Channel A2, prod. Brigitte Derenne and Robert Descharnes. Salvador Dali, Stéphane Lupasco, and André Robinet appeared on the show.

25. *Le Monde* (February 12–13, 1978): 12.

26. Salvador Dali, "Mystical Manifesto," in *The Collected Writings of Salvador Dali*, ed. and trans. Haim Finkelstein (Cambridge: Cambridge University Press, 1998), 216–19.

27. Bruno Froissart, "Salvador Dali et le monde angélique," appendix to *Journal d'un génie*, Salvador Dali (Paris: Gallimard, 1974), 309–12.

28. Quoted in Carme Ruiz, "Salvador Dali and Science," *El Punt*, October 18, 2000, Centre for Dalinian Studies, Fundació Gala–Salvador Dali, http://www.salvador-dali.org/serveis/ced/articles/en_article3.html.

29. "Dali and Science: Discover More," National Gallery of Victoria, n.d., http://www.ngv.vic.gov.au/dali/salvador/resources/daliandscience.pdf.

30. Ghyka, *Le nombre d'or*.

31. Matila Ghyka, *The Geometry of Art and Life* (New York: Sheed and Ward, 1946).

32. Matila Ghyka, *Esthétique des proportions dans la nature et dans les arts* (Paris: Gallimard, 1927).

33. David Lomas, "'Painting is dead—long live painting': Notes on Dali and Leonardo," *Papers of Surrealism* 4 (2006), http://www.surrealismcentre.ac.uk/papersofsurrealism/journal4/acrobat%20files/Lomaspdf.pdf.

34. Lomas, "'Painting is dead.'"

35. Salvador Dali, *50 Secrets of Magic Craftsmanship*, trans. Haakon Chevalier (New York: Dial Press, 1948).

36. As a proof of his admiration for Lupasco, Dali invited him to be part of the Honor Committee for the Membership Sword of the Academy of Fine Arts,

together with Giorgio de Chirico, Eugène Ionesco, and Félix Labisse: see Salvador Dalí to Stéphane Lupasco, letter, October 16, 1978, archives Alde Lupasco-Massot.

37. Salvador Dalí, "Gala, Velásquez and the Golden Fleece" (May 9, 1979), in *Dalí: L'oeuvre et l'homme*, Robert Descharnes (Lausanne: Edita, 1984).

38. This painting is in the possession of the Gala-Salvador Dalí Foundation in Figueres, Spain.

39. This painting is in the possession of the Gala-Salvador Dalí Foundation.

40. Ruiz, "Salvador Dalí and Science."

41. Salvador Dalí, *Oui: Méthode paranoïaque—critique et autres textes* (Paris: Denoël/Gonthier, 1971), 31.

42. Stéphane Lupasco, *Quelques considérations générales sur la peinture "abstraite" à propos des toiles de Benrath* (Mazamet: Babel Editeur, 1985). Twenty copies of this were illustrated with an acvaforte of Benrath; 200 copies contained a reproduction of one of Benrath's paintings. The original was printed in the catalog of the Frédéric Benrath exhibition in Gallery Prismes, Paris, in 1956.

43. Stéphane Lupasco, "Appel, Painter of Life," *XXe Siècle* 17 (1961). This issue of the magazine had a cover by Marc Chagall and an original lithography by Karel Appel. The text was republished in the catalog of Karel Appel's exposition *Reliefs 1966–1968* (Paris: CNAC, 1968).

44. Lupasco, *Quelques considérations*.

45. Lupasco, "Appel, Painter of Life" (Supplement "Chroniques du jour"): 1.

46. Karel Appel, "Le philosophe Stéphane Lupasco," 1956 (private collection), accessed January 13, 2006, http://www.bowi-groep.nl/index.php?module=gallerij&gallerij::artiestID=10.

47. René Huyghe, *Dialogue avec le visible* (Paris: Flammarion, 1955).

48. René Huyghe, *De l'art à la philosophie: Réponses à Simon Monneret* (Paris: Flammarion, 1980), 119, 156.

49. Lupasco, *Les trois matières*, 47, 151.

50. Gavin Parkinson, *Surrealism, Art and Modern Science: Relativity, Quantum Mechanics, Epistemology* (New Haven, CT: Yale University Press, 2008).

CHAPTER FIFTEEN

1. Henri Poincaré, "L'invention mathématique," *Bulletin de l'Institut général de Psychologie* 8, no. 3 (1908): 175–87; reprinted in Henri Poincaré, *Science et méthode* (Flammarion: Paris, 1908).

2. Jacques Hadamard, *An Essay on the Psychology of Invention in the Mathematical Field* (New York: Princeton University Press, 1945); French edition: *Essai sur la psychologie de l'invention dans le domaine mathématique* (Gauthier-Villars: Paris, 1978).

3. Poincaré, *Science et méthode*, 53–54.

4. Ibid., 54.

5. Ibid., 59.

6. Hadamard, *Essai sur la psychologie*, 19.

7. Ibid., 75.

8. Ibid., 77.

9. Ibid.

10. Ibid., 82.
11. Gilbert Durand, *Les Structures anthropologiques de l'imaginaire* (Bordas: Paris, 1981), 51.
12. James R. Newman, "Srinivasa Ramanujan," in *Mathematics in the Modern World: Readings from Scientific American*, ed. Morris Klein (San Francisco: W. H. Freeman and Co., 1968), 73–76.
13. Simon, *Kepler*, 186.
14. Ibid., 276–77.
15. J. Kepler, *L'Harmonie du monde*, trans. Jean Peyroux (Paris: Librairie Albert Blanchard, 1979).
16. Holton, *L'Imagination scientifique*, 83.
17. William James, *The Principles of Psychology* (1890; New York: Dover Publications, 1950).
18. Holton, *L'Imagination scientifique*, 103–104.
19. William James, quoted in Holton, *L'Imagination scientifique*, 110.
20. Niels Bohr, *La Théorie atomique et la description des phénomènes* (Paris: Gauthier-Villars, Paris, 1932), 108.
21. Holton, *L'Imagination scientifique*, 119.
22. Henri Michaux, *Connaissance par les gouffres* (Paris: Gallimard, 1972).
23. René Daumal, "Une Expérience fondamentale," in *Chaque Fois que l'Aube paraît*, René Daumal (Paris: Gallimard, 1953).
24. Henry Corbin, "Mundus imaginalis ou l'imaginaire et l'imaginal," in *Face de Dieu, face de l'homme: Herméneutique et soufisme*, Henri Corbin (Paris: Flammarion, 1983).
25. Gilbert Durand, "La reconquête de l'Imaginal," in *Henry Corbin*, ed. Christian Jambet, Cahier de l'Herne 39 (Paris: Editions de l'Herne, 1981), 269.

CHAPTER SIXTEEN

1. Guénon, *Le Règne de la quantité*.
2. Michel Henry, *La Barbarie* (Paris: Grasset, 1987).
3. Henri-Charles Puech, *En Quête de la Gnose, Vol. 2, Sur L'Evangile selon Thomas* (Paris: Gallimard, 1978), 23, logion 80.
4. Alan D. Sokal, "Transgressing the Boundaries: Toward a Transformative Hermeneutics of Quantum Gravity," *Social Text* 46/47 (1996): 336–61.
5. Alan D. Sokal, "A Physicist Experiments with Cultural Studies," *Lingua Franca* 6, no. 4 (1996): 62–64.
6. Steven Weinberg, "Sokal's Hoax," *New York Review of Books* 43, no. 13 (August 8, 1996).
7. Alan Sokal and Jean Bricmont, *Impostures intellectuelles* (Paris: Odile Jacob, 1997).
8. Ibid., 27.
9. Alan Sokal, *Pseudosciences et postmodernisme: Adversaires ou compagnons de route?* trans. Barbara Hochstedt (Paris: Odile Jacob, 2005), foreword by Jean Bricmont.
10. Ibid., 51, 155.
11. Ibid., 158.

12. Steven Weinberg, *Dreams of a Final Theory* (New York: Pantheon Books, 1992).
13. Dante Alighieri, "Le Paradis," *La Divine Comédie*, chant 28.
14. John Horgan, *The End of Science* (New York: Broadway Books, 1997).
15. Gablik, *Has Modernism Failed?*
16. Wolfgang Pauli, *Physique moderne et philosophie*, trans. Claude Maillard (Paris: Albin Michel, 1999), 178. Chapter 18, "Science and Western Thinking" (163–78), was first published in Wolfgang Pauli, *Europa: Erbe und Aufgabe* (Meinz: Internazionaler Gelehrtehkongress, 1955).
17. Nicolescu, *Manifesto of Transdisciplinarity*.
18. Basarab Nicolescu, "Transdisciplinarity as Methodological Framework for Going beyond the Science and Religion Debate," *Transdisciplinarity in Science and Religion* 2 (2007): 35–60.
19. Basarab Nicolescu, "Toward a Methodological Foundation of the Dialogue between the Technoscientific and Spiritual Cultures," in *Differentiation and Integration of Worldviews*, ed. Liubava Moreva (Saint Petersburg: Eidos, 2004).
20. Hans-Georg Gadamer, *Gesammelte Werke, Hermeneutik 1, Wahreit und Methode* (Tübingen: J. C. B. Mohr, 1960).

CHAPTER SEVENTEEN

1. "Reality," dictionary.com, 2013, http://dictionary.reference.com/browse/reality.
2. Roberto Poli, "The Basic Problem of the Theory of Levels of Reality," *Axiomathes* 12 (2001): 261–83; «Three Obstructions: Forms of Causation, Chronotopoids, and Levels of Reality," *Axiomathes* 1 (2007): 1–18.
3. Nicolai Hartmann, *Der Aufbau der realen Welt: Grundriss der allgemeinen Kategorienlehre* (Berlin: Walter De Gruyter, 1940).
4. Heisenberg, *Philosophie*.
5. Ibid., 240.
6. Ibid., 166.
7. Ibid., 145.
8. Ibid., 372.
9. Ibid., 235.
10. Ibid., 387.
11. Ibid., 261.
12. Nicolescu, *Manifesto of Transdisciplinarity*.
13. Paul Cilliers and Basarab Nicolescu, "Complexity and Transdisciplinarity: Discontinuity, Levels of Reality and the Hidden Third," *Futures* 44, no. 8 (2012): 711–18.
14. Edmund Husserl, *Méditations cartésiennes*, trans. Gabrielle Peiffer and Emmanuel Levinas (Paris: Vrin, 1966).
15. Joseph E. Brenner, *Logic in Reality* (New York: Springer, 2008).
16. Basarab Nicolescu, "The Idea of Levels of Reality and Its Relevance for Non-Reduction and Personhood," *Transdisciplinarity in Science and Religion* 4 (2008): 11–26.

17. Christopher C. Knight, *The God of Nature: Incarnation and Contemporary Science* (Minneapolis, MN: Fortress Press, 2007).
18. Basarab Nicolescu, *Théorèmes poétiques* (Monaco: Rocher, 1994).
19. Moraru, *Cosmodernism*.
20. Ibid., 8.
21. Ibid., 5.
22. Ibid., 7.
23. Ibid., 23, 57.
24. Ibid., 29.
25. Ibid., 55.
26. Ibid., 316.
27. Ibid., 304.
28. Ibid., 312.
29. Clara Janés, *La palabra y el secreto* (Madrid: Huerga & Fierro, 1999).
30. Clara Janés, "The Hidden Third," trans. Irina Dinca and Joseph Brenner, Pratique de la Transdisciplinarité, CIRET, http://ciret-transdisciplinarity.org/ARTICLES/liste_articles.php. Dedicated to Basarab Nicolescu. The Spanish original, "Tercero oculto," was published in *No tinguis res a les mans* (Sabadell, Spain: Papers de Versàlia, 2010), 35.
31. Basarab Nicolescu, "Towards an Apophatic Methodology of the Dialogue between Science and Religion," in *Science and Orthodoxy: A Necessary Dialogue*, ed. Basarab Nicolescu and Magda Stavinschi (Bucharest: Curtea Veche Publishing, 2006), 19–29.
32. Jesper Hoffmeyer, *Signs of Meaning in the Universe* (Bloomington: Indiana University Press, 1993).
33. *Biosemiotics in Transdisciplinary Contexts: Proceedings of the Gathering in Biosemiotics 6, Salzburg 2006*, ed. Günther Witzany (Helsinki: UMWEB Publications, 2007).
34. Charles Sanders Peirce, 4:383–84.

BIBLIOGRAPHY

Abbott, Edwin A. *Flatland: A Romance of Many Dimensions*. New York: Barnes & Noble, 2013.
Adonis. *La prière et l'épée: Essais sur la culture arabe*. Translated by Leïla Khatib and Anne Wade Minkowski. Paris: Mercure de France, 1993.
Alighieri, Dante. "Le Paradis." *La Divine Comédie*. Translated by Jacqueline Risset. Paris: Flammarion, 2004.
Aspect, Alain, Jean Dalibard, and Gérard Roger. Experimental Test of Bell's Inequalities Using Time-Varying Analyzers. *Physical Review Letters* 49, no. 25 (1982): 1804–807.
Aspect, Alain, Philippe Grangier, and Gérard Roger. Experimental Tests of Realistic Local Theories via Bell's Theorem. *Physical Review Letters* 47, no. 7 (1981): 460–63.
Atmanspacher, Harald, and Hans Primas. "Pauli's Ideas on Mind and Matter in the Context of Contemporary Science." *Journal of Consciousness Studies* 13, no. 3 (2006): 5–50.
Attar, Farid ud-Din. *The Conference of the Birds*. Translated by C. S. Nott. Boulder, CO: Shambhala, 1971.
Bădescu, Horia, and Basarab Nicolescu, eds. *Stéphane Lupasco: L'homme et l'oeuvre*. Monaco: Rocher, 1999.
Banu, Georges. "La Conférence des Oiseaux, ou le chemin vers soi-même." In *Les Voies de La Création Théâtrale*. Vol. 10, *Krejka-Brook*. Edited by Denis Bablet, 253–95. Paris: Editions de C.N.R.S., 1982.
Barrow, John D. *La Grande Théorie: Les limites d'une explication globale en physique*. Translated by Michel Cassé and Guy Paulus. Paris: Albin Michel, 1994.
Barrow, John, Frank Tipler, and Marie-Odile Monchicourt. *L'Homme et le cosmos: Le Principe anthropique en astrophysique moderne*. Paris: Imago, 1984.
Baruzi, Jean. *Saint Jean de la Croix et le problème de l'expérience mystique*. Paris: Librairie Félix Alcan, 1931.
Bell, J. S. *Speakable and Unspeakable in Quantum Mechanics: Collected Papers on Quantum Philosophy*. Cambridge: Cambridge University Press, 1987.
Berdiaeff, N. "L'Ungrund' et la liberté." In *Mysterium Magnum*. Vol. 1. Translated by N. Berdiaev, 6–45. Paris: Éditions d'Aujourd'hui, 1978.
Boehme, Jacob. *L'Aurore naissante*. Translated by Le Filosophe Inconnu. Milano: Arché, 1977.
———. "De Electione Gratiae." In *Jacob Boehme: Cahiers de l'Hermetisme*. Edited by Antoine Faivre and Frédérick Tristan, 68–70. Paris: Albin Michel, 1977.

———. "De la Base sublime et profonde des six points théosophiques." Translated by Louis-Claude de Saint-Martin. In *Jacob Boehme: Cahiers de l'Hermetisme*. Edited by Antoine Faivre and Frédérick Tristan, 115–86. Paris: Albin Michel, 1977.

———. *Mysterium Magnum*. Translated by N. Berdiaev. 4 vols. Paris: Éditions d'Aujourd'hui, 1978.

———. "Mysterium pansophicum." Translated by Louis-Claude de Saint-Martin. In *Jacob Boehme: Cahiers de l'Hermetisme*. Edited by Antoine Faivre and Frédérick Tristan, 187–98. Paris: Albin Michel, 1977.

Bohr, Niels. *Essays 1958–1962 on Atomic Physics and Human Knowledge*. New York: Interscience Publishers, 1963.

———. *La Théorie atomique et la description des phénomènes*. Paris: Gauthier-Villars, Paris, 1932.

Bosquet, Alain. *Conversations with Dali*. New York: E. P. Dutton & Co., 1969.

Brenner, Joseph E. *Logic in Reality*. New York: Springer, 2008.

Breton, André. *Anthologie de l'humour noir*. Paris: Sagittaire, 1950.

———. "Les artistes modernes se soucient moins de beauté que de liberté." Interview by André Parinaud. *Arts*, Paris, March 7, 1952. In *Entretiens (1913–1952)*. André Breton, 293–99. Paris: Gallimard, 1952.

———. Interview by Jose M. Valverde. *Correo literario*, Madrid, September 1950. In *Entretiens (1913–1952)*. André Breton, 281–85. Paris: Gallimard, 1952.

Brody, T. A. "On Quantum Logic." *Foundations of Physics* 14, no. 5 (1984): 409–30.

Brook, Peter. *The Empty Space*. Harmondsworth, UK: Penguin Books, 1977.

———. Interview by Ronald Hayman. *The Times*, August 29, 1970.

Calder, Nigel. *The Key of the Universe*. London: British Broadcasting Corporation, 1977.

Camus, Michel. "The Hidden Hand between Poetry and Science." In *Transdisciplinarity: Theory and Practice*. Edited by Basarab Nicolescu, 53–65. Creskill, NJ: Hampton Press, 2008.

Camus, Michel, Thierry Magnin, Basarab Nicolescu, and Karen-Claire Voss. "Levels of Representation and Levels of Reality: Towards an Ontology of Science." In *The Concept of Nature in Science and Theology*. Part 2. Edited by Niels H. Gregersen, Michael W. S. Parsons, and Christoph Wassermann, 94–103. Geneva: Labor et Fides, 1998.

Capra, Fritjof. *The Tao of Physics: An Exploration of the Parallels between Modern Physics and Eastern Mysticism*. Berkeley: Shambala, 1975.

Caro, Manuel J., and John W. Murphy, eds. *The World of Quantum Culture*. Westport, CT: Praeger Publishers, 2002.

Cazaban, Costin. "Temps musical/espace musical comme fonctions logiques." In *L'esprit de la musique: Essais d'esthétique et de philosophie*. Edited by Hugues Dufourt, Joël-Marie Fouquet, and François Hurard, 301–14. Paris: Klincksieck, 1992.

Cazenave, Michel. "Psychologie et tiers inclus." In *À la confluence de deux cultures: Lupasco aujourd'hui*. Edited by Basarab Nicolescu, 130–34. Paris: Oxus, 2010.

Cazenave, Michel. "La place et la fonction du transculturel dans le monde contemporain." *Transdisciplinary Studies: Science, Spirituality, Society* 1 (2011): 173–83.

Chew, Geoffrey F. "'Bootstrap': A Scientific Idea?" *Science* 161 (1968): 762–65.

———. "Hadron 'Bootstrap': Triumph or Frustration?" *Physics Today* 23, no. 10 (1970): 23–28.

---. "Impasse for the Elementary-Particle Concept." In *The Great Ideas Today 1974*. 114–19. London: Encyclopaedia Britannica, 1974.
---. *S-Matrix Theory of Strong Interaction*. New York: W. A. Benjamin, 1961.
---. "Theory of Strong Coupling of Ordinary Particles." In *Proceedings of the Ninth International Annual Conference on High Energy Physics*. Kiev, 1959.
Chew, G. F., and S. Mandelstam. "Theory of the Low-Energy Pion-Pion Interaction: II." *Nuovo Cimento* 19 (1961): 752–76.
Chirila, Ioan. "Ternaire et Trinité, homogène, hétérogène et l'*état T*': Une évaluation théologique du discours lupascien sur le tiers inclus." In *À la confluence de deux cultures: Lupasco aujourd'hui*. Edited by Basarab Nicolescu, 54–78. Paris: Oxus, 2010.
Cilliers, Paul. *Complexity and Postmodernism: Understanding Complex Systems*. London: Routledge, 1998.
Cilliers, Paul, and Basarab Nicolescu. "Complexity and Transdisciplinarity: Discontinuity, Levels of Reality and the Hidden Third." *Futures* 44, no. 8 (2012): 711–18.
Cioroianu, Adrian. "Stéphane Lupasco temps et contradiction: Vers une nouvelle logique de l'histoire?" In *À la confluence de deux cultures: Lupasco aujourd'hui*. Edited by Basarab Nicolescu, 20–39. Paris: Oxus, 2010.
CIRET: Centre International de Recherches et Études Transdisciplinaires. "Venice Declaration: Final Communiqué of the Symposium Science and the Boundaries of Knowledge: The Prologue of Our Cultural Past." March 7, 1986. Last modified June 1994. Accessed March 25, 2013. http://ciret-transdisciplinarity.org/bulletin/b2c4_en.php.
Clair, Jean. "Ouspensky et l'espace néo-platonicien." In *Malevitch*. Edited by Jean-Claude Marcadé, 15–29. Lausanne: L'Age d'Homme, 1979.
Cohen, C. *La femme des origines: Images de la femme dans la préhistoire occidentale*. Paris: Belin-Herscher, 2003.
Connes, Alain, André Lichnerowicz, and Marcel Paul Schützenberger. *Triangle de pensées*. Paris: Odile Jacob, 2000.
Corbin, Henry. *Face de Dieu, face de l'homme: Herméneutique et soufisme*. Paris: Flammarion, 1983.
---. "Mundus imaginalis ou l'imaginaire et l'imaginal." *Cahiers internationaux de symbolisme* 6 (1964): 3–26. Republished in *Face de Dieu, face de l'homme: Herméneutique et soufisme*. Henri Corbin. Paris: Flammarion, 1983.
Costache, Doru. "Logos, Evolution, and Finality in Anthropological Research: Towards a Transdisciplinary Solution." In *Science and Religion: Antagonism or Complementarity?* Edited by Basarab Nicolescu and Magda Stavinschi, 241–60. Bucharest: XXI Eonul dogmatic, 2003.
Couglan, G. D., and J. E. Dodd. *The Ideas of Particle Physics*. Cambridge: Cambridge University Press, 1991.
Craciunescu, Pompiliu. *Vintila Horia: Translittérature et Réalité*. Veauche, France: L'Homme Indivis, 2008.
Crafton, Anthony. *Cardano's Cosmos: The Worlds and Works of a Renaissance Astrologer*. Cambridge, MA: Harvard University Press, 1999.
Cushing, James T. "Is There Just One Possible World? Contingency vs. the Bootstrap." *Studies in the History and Philosophy of Science* 16 (1985): 31–48.

———. *Theory Construction and Selection in Modern Physics: The S Matrix.* Cambridge: Cambridge University Press, 1990.
Dali, Salvador. *Anti-Matter Manifesto.* Exhibition booklet. New York: Carstairs Gallery, 1958.
———. *50 Secrets of Magic Craftsmanship.* Translated by Haakon Chevalier. New York: Dial Press, 1948.
———. "Gala, Velásquez and the Golden Fleece." In *Dali: L'oeuvre et l'homme.* Robert Descharnes. Lausanne: Edita, 1984.
———. "Mystical Manifesto." In *The Collected Writings of Salvador Dali.* Edited and Translated by Haim Finkelstein. Cambridge: Cambridge University Press, 1998.
———. *Oui: Méthode paranoïaque—critique et autres textes.* Paris: Denoël/Gonthier, 1971.
———. "Reconstitution du corps glorieux dans le ciel." *Etudes Carmélitaines* 2 (1952): 171–72.
Damasio, Antonio R. "Die Insel des Lachens: Erstes Interview mit Antonio Damasio in Iowa City." Interview by Joscha Remus. *Infonautik: Wissen in Bewegung.* http://www.infonautik.de/damasio.htm.
———. *Looking for Spinoza: Joy, Sorrow, and the Feeling Brain.* San Diego, CA: Harcourt, 2003.
Daumal, René. "Une Expérience fondamentale." In *Chaque Fois que l'Aube paraît.* René Daumal, 265–74. Paris: Gallimard, 1953.
Davies, Paul. *Superforce: The Search for a Great Unified Theory of Nature.* New York: Simon & Schuster, 1984.
De Caso, Michel. "La mise à jour des Lumières: Tiers inclus, niveaux de Réalité et Rectoversion." In *À la confluence de deux cultures: Lupasco aujourd'hui.* Edited by Basarab Nicolescu, 228–49. Paris: Oxus, 2010.
de Gaigneron, Ludovic. *L'Image ou le drame de la nullité cosmique.* Paris: Le Cercle du Livre, 1956.
Descombes, Vincent. *Le complément du sujet.* Paris: Gallimard, 2004.
d'Espagnat, Bernard. *À la Recherche du Réel.* Paris: Gauthier-Villars, 1981.
———. "Nonseparability and the Tentative Descriptions of Reality." *Physics Reports* 110, no. 4 (1987): 201–64.
———. "The Quantum Theory and Reality." *Scientific American* 241, no. 5 (1979): 128–40.
Dicke, R. H. "Dirac's Cosmology and Mach's Principle." *Nature* 192 (1961): 440–41.
Dirac, P. A. M. "The Cosmological Constants." *Nature* 139 (1937): 351–52.
Dobbs, B. J. T. *The Janus Faces of Genius: The Role of Alchemy in Newton's Thought.* Cambridge: Cambridge University Press, 1991.
Dumont, Jean-Paul. "Anaxagore." In *Dictionnaire des philosophes.* Edited by Denis Huisman, 77–81. Paris: PUF, 1984.
Durand, Gilbert. "L'Anthropologie et les structures du complexe." In *Stéphane Lupasco: L'homme et l'oeuvre.* Edited by Horia Badescu and Basarab Nicolescu, 61–74. Monaco: Rocher, 1999.
———. *L'Imagination symbolique.* Paris: Quadrige/PUF, 1984.
———. "La reconquête de l'Imaginal." In *Henry Corbin.* Edited by Christian Jambet, 266–73. Cahier de l'Herne 39. Paris: Editions de l'Herne, 1981.

———. *Les Structures anthropologiques de l'imaginaire*. Bordas: Paris, 1981.
Dürrenmatt, Friedrich. *Albert Einstein*. Translated by Françoise Stokke. Lausanne: Éditions de l'Aire, 1982.
Eccles, John. *How the Self Controls Its Brain*. New York: Springer Verlag, 1994.
Eco, Umberto. *Les limites de l'interprétation*. Paris: Grasset, 1992.
Eddington, A. S. *La Nature du monde physique*. Paris: Payot, 1929.
———. *The Philosophy of Physical Science*. Cambridge: Cambridge University Press, 1949. First published 1939 by Macmillan.
Einstein, Albert. *Comment je vois le Monde*. Paris: Flammarion, 1934.
———. "Remarques préliminaires sur les concepts fondamentaux." In *Louis de Broglie: Physicien et penseur*. Albert Einstein, 5–15. Paris: Albin Michel, 1953.
Eisele, C., ed. *Charles Sanders Peirce: The New Elements of Mathematics*. 4 vols. The Hague: Mouton Humanities Press, 1976.
Eliade, Mircea. *Images and Symbols: Studies in Religious Symbolism*. Translated by Philip Mairet. Princeton: Princeton University Press, 1991.
———. *Ordeal by Labyrinth: Conversations with Claude-Henri Roquet*. Translated by Derek Cotman Chicago: University of Chicago Press, 1983.
Finkelstein, D. "Holistic Methods in Quantum Logic." In *Quantum Theory and the Structure of Time and Space*. Vol. 3. Edited by L. Castell and M. Drieschner. Munich: Carl Hanser Verlag, 1979.
Fischer, Carol Anne. "Quantum Theatre: A Language for the Voices of Contemporary Theater." Master's thesis, San Jose State University, May 2004. http://www.tvradiofilmtheatre.org/MA/Pages/thesesonline/FischerTHESIS.html.
Frank, Philippe. *Einstein: Sa vie et son temps*. Paris: Albin Michel, 1950.
Freedman, D. Z., and P. van Nieuwenhuizen. "La Supergravité et l'unification des lois de la physique." In *Les Particules élémentaires*, 193–206. Paris: Bibliothèque "Pour la science," 1983.
Froissart, Bruno. "Salvador Dali et le monde angélique." In *Journal d'un génie*. Salvador Dali, 309–12. Paris: Gallimard, 1974.
Gablik, Suzi. *Has Modernism Failed?* New York: Thames & Hudson, 2004.
Gadamer, Hans-Georg. *Gesammelte Werke. Hermeneutik 1. Wahreit und Methode*. Tübingen: J. C. B. Mohr, 1960.
Gale, George. "Chew's Monadology." *History of Ideas* 35 (1974): 339–48.
———. "Leibniz, Chew and Wheeler on the Identity of Physical and Philosophical Inquiry." *Review of Metaphysics* 29 (1975): 323–33.
Galilei, Galileo. *Dialogue Concerning the Two Chief World Systems, Ptolemaic and Copernican*. Translated by Stillman Drake. Berkeley: University of California Press, 1962.
Gell-Mann, Murray. *Le Quark et le jaguar*. Translated by Gilles Minot. Paris: Flammarion, 1997.
Ghils, Paul. "Langage pur, langage impur: Du mythe de l'origine à la pragmatique de la contradiction." In *À la confluence de deux cultures: Lupasco aujourd'hui*. Edited by Basarab Nicolescu, 159–96. Paris: Oxus, 2010.
Ghyka, Matila. *Esthétique des proportions dans la nature et dans les arts*. Paris: Gallimard, 1927.
———. *The Geometry of Art and Life*. New York: Sheed and Ward, 1946.

———. *Le nombre d'or: Rites et rythmes pythagoriciens dans le development de la civilisation occidentale*. Paris: Gallimard, 1931.
Glashow, Sheldon L. *Le Charme de la physique*. Translated by Olivier Colardelle. Paris: Albin Michel, 1997.
Goldman, Steven Louis. "Alexander Kojève on the Origin of Modern Science: Sociological Modelling Gone Awry." *Studies in History and Philosophy of Science* 6, no. 2 (1975): 113–24.
Graisser, Julia H., and T. K. Graisser. "Partons in Antiquity." *American Journal of Physics* 45, no. 5 (1977): 439–42.
Granger, Gilles Gaston. *L'irrationnel*. Paris: Odile Jacob, 1998.
Greene, Brian. *The Elegant Universe: Superstrings, Hidden Dimensions, and the Quest for the Ultimate Theory*. London: Vintage Series, Random House, Inc., 2000.
Gross, David. "On the Uniqueness of Field Theories." In *A Passion for Physics: Proceedings of the G. F. Chew Jubilee, September 29, 1984*. Edited by Carleton DeTar, J. Finkelstein, and Chung-I Tan, 128–36. Singapore: World Scientific, 1985.
Guénon, René. *La Grande Triade*. Paris: Gallimard, 1974.
———. *Le Règne de la quantité et les signes des temps*. Paris: Gallimard, 1970.
———. *Le Symbolisme de la Croix*. Paris: Vega, 1970.
Guth, A. H. "Inflationary Universe: A Possible Solution to the Horizon and Flatness Problems." *The Physical Review* D23, no. 2 (1981): 347–56.
Hadamard, Jacques. *An Essay on the Psychology of Invention in the Mathematical Field*. New York: Princeton University Press, 1945.
Harambourg, Lydia. *Georges Mathieu*. Neuchâtel: Ides et Calandes, 2001.
Hartmann, Nicolai. *Der Aufbau der realen Welt: Grundriss der allgemeinen Kategorienlehre*. Berlin: Walter De Gruyter, 1940.
Hawking, Stephen. "Les Objectifs de la physique théorique." *CERN Courier* 21 (January-February 1981): 3–7.
Heilpern, John. *Conference of the Birds: The Story of Peter Brook in Africa*. Harmondsworth, UK: Penguin, 1979.
Heisenberg, Werner. *Philosophie: Le manuscrit de 1942*. Translated by Catherine Chevalley. Paris: Seuil, 1998. Originally published in German as *Ordnung der Wirklichkeit*. In *W. Heisenberg Gesammelte Werke*. Vol. C-I. *Physik und Erkenntnis, 1927–1955*. Edited by W. Blum, H. P. Dürr, H. Rechenberg, and R. Piper, 218–306. Munich: GmbH & KG, 1984.
Henderson, Linda Dalrymple. *The Fourth Dimension and Non-Euclidean Geometry in Modern Art*. Princeton, NJ: Princeton University Press, 1983.
———. "The Merging of Time and Space: The Fourth Dimension in Russia from Ouspensky to Malevitch." *The Structurist* 15/16 (1975/1976): 97–108.
Henry, Michel. *La Barbarie*. Paris: Grasset, 1987.
Hinton, Charles Howard. *The Fourth Dimension*. Salem, NH: Ayer Company, 1986. Reproduction of the 1912 edition. London: Allen & Unwin Ltd.
———. *Scientific Romances*. 2nd series. Leeds: Celephaïs Press, 2008. Available at http://fr.scribd.com/doc/8552379/Hinton-Scientific-Romances-2. First published in 1896 by Swan Sonnenschien.
Hoffmeyer, Jesper. *Signs of Meaning in the Universe*. Bloomington: Indiana University Press, 1993.

Holton, Gerald. *Thematic Origin of Scientific Thought: From Kepler to Einstein.* Cambridge, MA: Harvard University Press, 1973.
———. *The Scientific Imagination: Case Studies.* Cambridge: Cambridge University Press, 1978.
Horgan, John. *The End of Science.* New York: Broadway Books, 1997.
Horia, Vintila. *Viaje a los centres de la Tierra: Encuesta sobre el Estado Actual del Pensamiento, Las Artes y Las Ciencias.* Barcelona: Ediciones de Nuevo Arte Thor, 1987.
Husserl, Edmund. *Cartesian Meditations: An Introduction to Phenomenology.* Dordrecht: Kluwer Academic Publishers, 1977.
Huyghe, René. *De l'art à la philosophie: Réponses à Simon Monneret.* Paris: Flammarion, 1980.
———. *Dialogue avec le visible.* Paris: Flammarion, 1955.
Ionesco, Eugène. *Journal en miettes.* Paris: Mercure de France, 1967.
———. *Théâtre complet.* Edited by Emmanuel Jacquart. Pléiade series. Paris: Gallimard, 1990.
———. *Victimes du devoir.* In *Théâtre: Tome I.* Paris: Gallimard, 1984. First published in 1954.
Jacob, Maurice, and Geoffrey F. Chew. *Strong-Interaction Physics: A Lecture Note Volume.* New York: W. A. Benjamin, 1964.
James, William. *The Principles of Psychology.* New York: Dover Publications, 1950. First published in 1890.
Jammer, Max. *The Conceptual Development of Quantum Mechanics.* New York: McGraw-Hill Book Company, 1966.
Janés, Clara. "The Hidden Third." Translated by Irina Dinca and Joseph Brenner. Pratique de la Transdisciplinarité, CIRET. http://ciret-transdisciplinarity.org/ARTICLES/liste_articles.php.
———. *La palabra y el secreto.* Madrid: Huerga & Fierro, 1999.
Jantsch, Erich. *The Self-Organizing Universe.* Oxford: Pergamon Press, 1980.
Jean-Blain, Marguerite. *Eugène Ionesco: Mystique ou mal-croyant?* Brussels: Lessius, 2005.
Saint John of the Cross. *The Complete Works.* Vol. 1. Charleston, SC: Nabu Press, 2010.
———. "La Montée du Carmel." In *Oeuvres spirituelles.* Translated by R. P. Grégoire de Saint Joseph, 15–472. Paris: Seuil, 1954.
Juarroz, Roberto. *Nouvelle poésie verticale.* Paris: Lettres Vives, 1984.
Jung, C. G. *Commentaire sur le Mystère de la Fleur d'Or.* Paris: Albin Michel, 1980.
———. *Essais sur la symbolique de l'esprit.* Paris: Albin Michel, 1991.
———. *Ma vie: Souvenirs, rêves et pensées.* Edited by Aniela Jaffé. Paris: Gallimard, 1992.
Kaku, Michio. *Hyperspace: A Scientific Odyssey through Parallel Universes, Time Warps, and the Tenth Dimension.* Oxford: Oxford University Press, 1994.
Kepler, Johannes. *The Harmony of the World.* Translated by Eric J. Aiton, Alistair Matheson Duncan, and Judith Veronica Field. New York: American Philosophical Society, 1997.
King, Elliott H. "Nuclear mysticism." In *Salvador Dali: Liquid Desire.* Melbourne: National Gallery of Victoria, 2009.

Knight, Christopher C. *The God of Nature: Incarnation and Contemporary Science*. Minneapolis, MN: Fortress Press, 2007.
Kojève, Alexandre. "L'origine chrétienne de la science moderne." In *Mélanges Alexandre Koyré*. Vol. 2. *L'aventure de l'esprit*. Edited by Fernand Braudel, 295–306. Paris: Hermann, 1964.
Korzybski, Alfred. *Science and Sanity*. Lakeville, CT: The International Non-Aristotelian Library Publishing Company, 1958.
Kremer-Marietti, Angèle. "Gerald Holton." In *Dictionnaire des philosophes*. Vol. 1. Edited by Denis Huisman, 1247–51. Paris: PUF, 1984.
Kustow, Michael. *Peter Brook: A Biography*. New York: St. Martin's Press, 2005.
Laurikainen, K. V. *Beyond the Atom: The Philosophical Thought of Wolfgang Pauli*. Berlin: Springer, 1988.
Laszlo, Ervin. *Le Systémisme: Vision nouvelle du monde*. Paris: Pergamon Press, 1981.
Lederman, Leon. *The God Particle: If the Universe Is the Answer, What Is the Question?* Boston: Houghton Mifflin Company, 1993.
Leupin, Alexandre. *Fiction et Incarnation: Littérature et théologie au Moyen Age*. Paris: Flammarion, 1993.
Lomas, David. "'Painting is dead—long live painting': Notes on Dali and Leonardo." *Papers of Surrealism* 4 (2006). http://www.surrealismcentre.ac.uk/papersofsurrealism/journal4/acrobat%20files/Lomaspdf.pdf.
Lupasco, Stéphane. "Appel, Painter of Life." *XXe Siècle* 17, supp. Chroniques du jour (1961): 1–2.
———. "Dali and Sub-Realism." *XXe Siècle*, special issue Homage to Salvador Dali (1980): 117–18.
———. *Du devenir logique et de l'affectivité*. 2 vols. Paris: Vrin, 1973.
———. *L'expérience microphysique et la pensée humaine*. 2nd ed. Monaco: Rocher, 1989.
———. "Le principe d'antagonisme et l'art abstrait." *Ring des Arts*, no. 1 (1960): 6–21.
———. *Le principe d'antagonisme et la logique de l'énergie: Prolégomènes à une science de la contradiction*. Paris: Hermann, 1951.
———. *Psychisme et sociologie*. Paris: Casterman, 1978.
———. "Quelques aperçus sur la logique dynamique du contradictoire." In *United States Lines: Paris Review*. Edited by Georges Mathieu. Tours: Arrault et Cie, 1953.
———. *Quelques considérations générales sur la peinture "abstraite" à propos des toiles de Benrath*. Mazamet: Babel Editeur, 1985.
———. "Réponse au questionnaire d'André Breton." In *L'art magique*. André Breton with G. Legrand, 76–77. Paris: Club français du livre, 1957.
———. *Les trois matières*. Paris: Julliard, 1970.
Magnin, Thierry. *Entre science et religion: Quête de sens dans le monde présent*. Monaco: Rocher, 1998.
———. "L'unité des antagonismes dans l'histoire de la théologie catholique." In *À la confluence de deux cultures: Lupasco aujourd'hui*. Edited by Basarab Nicolescu, 79–102. Paris: Oxus, 2010.
Maillard, Christine. *Les sept sermons aux morts de Carl Gustav Jung*. Nancy: Presses Universitaires de Nancy, 1993.
Maki, Z., Y. Ohnuki, and S. Sakata. "Remarks on a New Concept of Elementary Particles and the Method of the Composite Model." *Proceedings of the Inter-

national Conference on Elementary Particles. Edited by Y. Tanikawa, 109–15. Kyoto: Kyoto University, 1965.

Malherbe, Jean-François. "Un choix décisif à l'aube de l'éthique: Parménide ou Héraclite." Transdisciplinarity in Science and Religion 5 (2009): 11–36.

———. "Esquisse d'une histoire de l'éthique à l'aune du tiers inclus." In À la confluence de deux cultures: Lupasco aujourd'hui. Edited by Basarab Nicolescu, 40–53. Paris: Oxus, 2010.

———. "Jeux de langage" et "tiers inclus": De nouveaux outils pour l'éthique appliquée. Sherbrooke, QC: Université de Sherbrooke, GGC Éditions, 2000.

———. Le Nomade polyglotte. Saint-Laurent, QC: Bellarmin, 2000.

Marcus, Solomon. Introduction mathématique à la linguistique structurale. Paris: Dunod, 1967.

Mathieu, Georges. L'abstraction prophétique. Paris: Gallimard, 1984.

———. Au-delà du Tachisme. Paris: Julliard, 1963.

———. Mathieu: 50 ans de création. Paris: Hervas, 2003.

———. "Mon ami Lupasco." Lecture presented at the International Congress "Stéphane Lupasco: L'homme et l'oeuvre," March 13, 1998, Académie des Inscriptions et Belles Lettres, Paris. In Stéphane Lupasco: L'homme et l'oeuvre, ed. Horia Bădescu and Basarab Nicolescu, 13–28. Monaco: Rocher, 1999.

Menta, Ed. The Magic World behind the Curtain: Andrei Serban in the American Theater. New York: Peter Lang, 1995.

Merlo, J. P. "Physique et philosophie." Preprint, submitted 1975. Centre d'Études Nucléaires de Saclay.

Michaux, Henri. Connaissance par les gouffres. Paris: Gallimard, 1972.

Minkowski, Hermann. "Raum und Zeit." Paper presented at the Eightieth Meeting of German Physicists and Physicians, Köln, September 21, 1908. Physikalische Zeitschrift 10 (1909): 75–88.

Modreanu, Simona. "A Different Approach to the 'Theater of the Absurd' with Special Reference to Eugène Ionesco." Cultura: International Journal of Philosophy and Axiology 8, no. 1 (2011): 171–86.

Montassier, Gérard. Le Fait Culturel. Paris: Fayard, 1980.

Morales, Gregorio. El cadáver de Balzac: Una visión cuántica de la literatura y el arte. Alicante: Epígono,1998.

———. Canto cuántico. Granada: Dauro, 2003.

———. "Overcoming the Limit Syndrome." In The World of Quantum Culture. Edited by Manuel J. Caro and John W. Murphy, 1–34. Westport, CT: Praeger Publishers, 2002.

———. "Xaverio's Quantum Aesthetics." In Xaverio: Estética cuantica, petrales 1997–2000. Granada: Casa de Granda, 2000.

Moraru, Christian. Cosmodernism: American Narrative, Late Globalization, and the New Cultural Imaginary. Ann Arbor: The University of Michigan Press, 2011.

Morel, Bernard. Dialectiques du Mystère. Paris: La Colombe, 1962.

Morin, Edgar. "Lupasco et les pensées qui affrontent les contradictions." In À la confluence de deux cultures: Lupasco aujourd'hui. Edited by Basarab Nicolescu, 103–29. Paris: Oxus, 2010.

———. La Méthode. Vol. 1. La Nature de la Nature. Paris: Seuil, 1977.

———. *La Méthode*. Vol. 2. *La Vie de la Vie*. Paris: Seuil, 1980.
———. *La Méthode*. Vol. 3. *La Connaissance de la connaissance*. Paris: Seuil, 1986.
———. *Science avec conscience*. Paris: Fayard, 1982.
Morris, Michael S., Kip S. Thorne, and Ulvi Yurtsever. "Wormholes, Time Machines, and the Weak Energy Condition." *Physical Review Letters* 61, no. 13 (1988): 1446–449.
Nagel, Ernest, and James R. Newman. *Gödel's Proof*. New York: New York University Press, 1958.
Needham, Joseph. *Within the four seas: The dialogue of East and West*. London: Allen & Unwin Ltd, 1969.
Ne'eman, Yuval. "Concrete versus Abstract Theoretical Models." In *Interaction between Science and Philosophy*. Edited by Y. Elkana. Atlantic Highlands, NJ: Humanities Press, 1975.
Newman, James R. "Srinivasa Ramanujan." In *Mathematics in the Modern World: Readings from Scientific American*. Edited by Morris Klein. San Francisco: W. H. Freeman and Co., 1968.
Basarab Nicolescu, "Gödelian Aspects of Nature and Knowledge." In *Systems: New Paradigms for the Human Sciences*. Edited by Gabriel Altmann and Walter A. Koch, 385–403. Berlin: De Gruyter, 1998.
———. "The Idea of Levels of Reality and Its Relevance for Non-Reduction and Personhood." *Transdisciplinarity in Science and Religion* 4 (2008): 11–26.
———. *Manifesto of Transdisciplinarity*. Translated by Karen-Claire Voss. New York: State University of New York Press, 2002.
———. "La mécanique quantique, l'évolution et le progrès." *Question de* 118 (1999): 23–30.
———. *Nous, la particule et le monde*. 3rd ed. Brussels: E. M. E. & InterCommunications, 2012.
———. "Peter Brook and Traditional Thought." Translated by David Williams. *Contemporary Theatre Review* 7 (1997): 11–23.
———. "Quelques réflexions sur la pensée atomiste et la pensée systémique." 3e *Millénaire* 7 (1983): 20–26.
———. "Relativité et physique quantique." In *Dictionnaire de l'ignorance*. Edited by Michel Cazenave, 108–20. Paris: Albin Michel, 1998.
———. "Science as Testimony." In *Science and The Boundaries of Knowledge: The Prologue of Our Cultural Past*. Final Report REX-86/WS-23, 9–29. Paris: UNESCO, 1986.
———. *Science, Meaning, & Evolution: The Cosmology of Jacob Boehme*. Translated by Rob Baker. New York: Parabola Books, 1991.
———. "Stéphane Lupasco (1900–1988)." In *Universalia 1989*. Paris: Encyclopaedia Universalis, 1989.
———. "Stéphane Lupasco (1900–1988)." In *Les oeuvres philosophiques: Dictionnaire*. Vol. 2. Edited by Jean-François Mattéi, 3503. Paris: Encyclopédie Philosophique Universelle, 1992.
———. *Théorèmes poétiques*. Monaco: Rocher, 1994.
———. "Towards an Apophatic Methodology of the Dialogue between Science and Religion." In *Science and Orthodoxy: A Necessary Dialogue*. Edited by

Basarab Nicolescu and Magda Stavinschi, 19–29. Bucharest: Curtea Veche Publishing, 2006.

———. "Toward a Methodological Foundation of the Dialogue between the Techno-scientific and Spiritual Cultures." In *Differentiation and Integration of Worldviews*. Edited by Liubava Moreva, 139–152. Saint Petersburg: Eidos, 2004.

———. *La transdisciplinarité: Manifeste*. Monaco: Rocher, 1996.

———. "Transdisciplinarity as Methodological Framework for Going beyond the Science and Religion Debate." *Transdisciplinarity in Science and Religion* 2 (2007): 35–60.

Nottale, Laurent. *Fractal Space-Time and Microphysics: Towards a Theory of Scale Relativity*. Singapore: World Scientific, 1992.

Ouspensky, P. D. *Fragments d'un enseignement inconnu*. Translated by Philippe Lavastine. Paris: Stock, 1978.

———. *Tertium Organum: A Key to the Enigmas of the World*. Translated by Nicholas Bessaraboff and Claude Bragdon. Rochester, NY: Manas Press, 1920. Originally published in Russian. Saint Petersbourg: N. P. Taberio, 1911.

Pagels, Heinz R. *The Cosmic Code*. Toronto: Bantam Books, 1983.

Panikkar, Raimon. *The Intrareligious Dialogue*. Mahwah, NJ: Paulist Press, 1999.

Parkinson, Gavin. *Surrealism, Art and Modern Science: Relativity, Quantum Mechanics, Epistemology*. New Haven, CT: Yale University Press, 2008.

Pauli, Wolfgang. "Die Wissenschaft und das abendländische Denken" [Science and Western Thinking]. In *Europa: Erbe und Aufgabe*. Meinz: Internazionaler Gelehrtehkongress, 1955.

———. *Physique moderne et philosophie*. Translated by Claude Maillard. Paris: Albin Michel, 1999.

Peirce, Charles Sanders. *Écrits sur le signe*. Edited by Gérard Deledalle. Paris: Seuil, 1978.

———. *Values in a Universe of Chance: Selected Writings of Charles S. Peirce*. Edited by Philip P. Wiener. New York: Dover Publications, 1966.

Penrose, Roger. *Shadows of the Mind: A Search for the Missing Science of Consciousness*. Oxford: Oxford University Press, 1995.

Penrose, Roger, and M. A. H. McCollum. "Twistor Theory: An Approach to the Quantization of Fields and Space-Time." *Physics Reports* 6C, no. 4 (1973): 241–316.

Planck, Max. *Autobiographie scientifique*. Translated by André George. Paris: Albin Michel, 1960.

———. *Initiation à la physique*. Paris: Flammarion, 1941.

Poincaré, Henri. "L'invention mathématique." *Bulletin de l'Institut général de Psychologie* 8, no. 3 (1908): 175–87.

———. *Science and Method*. Mineola, NY: Dover, 2003. Originally published as *Science et méthode*. Flammarion: Paris, 1908.

Poli, Roberto. "The Basic Problem of the Theory of Levels of Reality." *Axiomathes* 12 (2001): 261–83.

———. "Three Obstructions: Forms of Causation, Chronotopoids, and Levels of Reality." *Axiomathes* 1 (2007): 1–18.

Polizzotti, Mark. *André Breton*. Paris: Gallimard, 1995.

Priestley, J. B. *Man and Time* (London: Aldus Books, 1964).
Puech, Henri Charles. *En Quête de la Gnose*. Vol. 2. *Sur L'Evangile selon Thomas*. Paris: Gallimard, 1978.
Régy, Claude. *Espaces perdus*. Besançon: Les Solitaires Intempestifs, 1998.
———. *L'état d'incertitude*. Besançon: Les Solitaires Intempestifs, 2002.
Rennie, Bryan S., ed., *Changing Religious Worlds: The Meaning and End of Mircea Eliade*. Albany, NY: State University of New York Press, 2001.
Revardel, Jean-Louis. "Lupasco et la translogique de l'affectivité." In *À la confluence de deux cultures: Lupasco aujourd'hui*. Edited by Basarab Nicolescu, 135–58. Paris: Oxus, 2010.
Riger-Brown, Allan. "Gregorio Morales' Quantum Song: A Poetic Voyage into the Enfolded Order." *ELVIRA: Revista de Estudios Filológicos* 10 (2005). Accessed June 4, 2013. http://gregoriomorales.eshost.es/quantum_song_review.htm.
Roberts, Don D. *The Existential Graphs of Charles S. Peirce*. The Hague: Mouton Humanities Press, 1973.
Rosensohn, W. L. *The Phenomenology of Charles S. Peirce: From the Doctrine of Categories to Phaneroscopy*. Amsterdam: B. R. Grüner, 1974.
Rostain, Michel. "Journal des répétitions de *La Tragédie de Carmen*." In *Les Voies de la Création Théâtrale*. Vol. 13. *Peter Brook*. Edited by Georges Banu, 191–218. Paris: Editions de C.N.R.S., 1985.
Rucker, Rudy. *La quatrième dimension*. Translated by Christian Jeanmougin. Paris: Seuil, 1985.
Ruiz, Carme. "Salvador Dali and Science." *El Punt*, October 18, 2000. Available at Centre for Dalinian Studies, Fundació Gala–Salvador Dali. http://www.salvador-dali.org/serveis/ced/articles/en_article3.html.
Rûmî, Djalâl-ud-Dîn. *Le livre du dedans (Fîhi-mâ-fîhi)*. Translated by Eva de Vitray-Meyerovitch. Paris: Sindbad, 1975.
Salam, Abdus. "The Nature of the 'Ultimate' Explanation in Physics." In *Scientific Explanation: Papers Based on Herbert Spencer Lectures Given in the University of Oxford*. Edited by A. F. Heath. Gloucestershire: Clarendon, 1981.
———. "L'Unité des quatre énergies de l'univers." Interview by Michel Random. *3e Millénaire* 2 (1982): 14–19.
Sarraute, Nathalie. *Pour un oui ou pour un non*. Paris: Gallimard, 1998.
Science and the Boundaries of Knowledge: The Prologue of Our Cultural Past. Final Report REX-86/WS-2. Paris: UNESCO, 1986.
Segré, Emilio. *Les Physiciens modernes et leurs découvertes*. Paris: Fayard, 1984.
Shlein, Leonard. *Art and Physics: Parallel Visions in Space, Time, and Light*. New York: Harper Perennial, 2007.
Silburn, Lilian. *Instant et cause: Le discontinu dans la pensée philosophique de l'Inde*. Paris: Vrin, 1995.
Simon, Gérard. *Kepler astronome astrologue*. Paris: Gallimard, 1979.
Simón, Pablo Iglesias. "Experiencias en torno al teatro cuántico desde el lado oeste del Golden Gate." *ADE Teatro* 132 (2010): 210–13.
Smith, A. G. H. *Orghast at Persepolis*. London: Eyre Methuen, 1972.
Smolin, Lee. *The Trouble with Physics: The Rise of String Theory, the Fall of a Science, and What Comes Next*. London: Penguin Books, 2006.

Snow, C. P. *The Two Cultures and the Scientific Revolution.* New York: Cambridge University Press, 1961.
Sokal, Alan D. "A Physicist Experiments with Cultural Studies." *Lingua Franca* 6, no. 4 (1996): 62–64.
———. *Pseudosciences et postmodernisme: Adversaires ou compagnons de route?* Translated by Barbara Hochstedt. Paris: Odile Jacob, 2005.
———. "Transgressing the Boundaries: Toward a Transformative Hermeneutics of Quantum Gravity." *Social Text* 46/47 (1996): 336–61.
Sokal, Alan, and Jean Bricmont. *Impostures intellectuelles.* Paris: Odile Jacob, 1997.
Stapp, Henry P. *Mind, Matter and Quantum Mechanics.* New York: Springer Verlag, 1993.
Stearns, Isabel. "Firstness, Secondness, and Thirdness." In *Studies in the Philosophy of Charles Sanders Peirce.* Edited by Philip P. Wiener and Frederic H. Young. Cambridge, MA: Harvard University Press, 1952.
Steiner, George. "Penser Europe." In *L'Europe en quête d'harmonie.* Edited by Aude Fonquernie, 42–68. Mazille, France: La Maison sur le Monde, 2001.
Susskind, Leonard. *The Cosmic Landscape: String Theory and the Illusion of Intelligent Design.* New York: Bay Books, 2006.
Sypher, Wylie. *Loss of the Self in Modern Literature and Art.* New York: Random House, 1962.
Tegmark, Max, and John Archibald Wheeler. "100 Years of the Quantum." *Scientific American* 2 (2001): 68–75.
Teilhard de Chardin, Pierre. *Oeuvres.* 10 vols. Paris: Seuil, 1965–1970.
Thirring, Walter. "Do the Laws of Nature Evolve?" In *What Is Life? The Next Fifty Years: Speculations on the Future of Biology.* Edited by Michael P. Murphy and Luke A. O'Neil, 131–36. Cambridge: Cambridge University Press, 1995.
Tipler, Frank J. *The Physics of Immortality: Modern Cosmology, God and the Resurrection of the Dead.* New York: Doubleday, 1994.
von Bertalanffy, Ludwig. *Théorie générale des systèmes.* Paris: Dunod, 1973.
von Franz, Marie-Louise. *Nombre et temps: Psychologie des profondeurs et physique moderne.* Paris: La Fontaine de Pierre, 1983.
von Meyenn, K., ed. *Wolfgang Pauli: Wissenschaftlicher Briefwechse.* 4 vols. Berlin: Springer, 1993.
von Weizsäcker, C. F. "A Reconstruction of Quantum Theory." In *Quantum Theory and the Structure of Time and Space.* Vol. 3. Edited by L. Castell and M. Drieschner. Munich: Carl Hanser Verlag, 1979.
Weinberg, Steven. *Dreams of a Final Theory.* New York: Pantheon Books, 1992.
———. "Sokal's Hoax." *New York Review of Books* 43, no. 13 (August 8, 1996).
———. *Les Trois Premières Minutes de l'univers.* Translated by Jean-Benoît Yelnik. Paris: Seuil, 1978.
———. "Unified Theories of Elementary-Particle Interaction." *Scientific American* 231, no. 1 (1974): 50–59.
Whittaker, E. *Eddington's Principle in the Philosophy of Science.* Cambridge: Cambridge University Press, 1951.
Witten, Edward. "Duality, Spacetime and Quantum Mechanics." *Physics Today* 50, no. 5 (May 1997): 28–33.

Witzany, Günther, ed. *Biosemiotics in Transdisciplinary Contexts: Proceedings of the Gathering in Biosemiotics 6, Salzburg 2006*. Helsinki: UMWEB Publications, 2007.

Wunenburger, Jean-Jacques. *La raison contradictoire: Sciences et philosophie modernes: La pensée du complexe*. Paris: Albin Michel, 1990.

Zeami, *La tradition secrète de No*. Paris: Gallimard, 1960.

———. *On the Art of the No Drama: The Major Treatises of Zeami*. Translated by J. Thomas Rimer and Yamazaki Masakazu. Princeton, NJ: Princeton University Press, 1984.

NAME INDEX

Abellio, Raymond, 170
Abbott, Edwin A., 71, 76, 222n19, 223n2, 239
Adonis, 16, 218n36, 239
Alighieri, Dante, 198, 236n13, 239
Alvard, Julien, 170, 175
Anaxagoras, 96–97, 102
Antheil, George, 81
Appel, Karel, vii, 174–75, 234n43, 234n45–46, 246
Aristotle, vii, 4, 38, 168–70
Aspect, Alain, 57, 221n16, 239
Saint Augustine, 14, 75
Atmanspacher, Harald, 141, 229n10, 239
Attar, Farid ud-Din, 22, 27, 45, 161, 219n4, 221n2, 239
  Conference of the Birds, 155, 219n4, 221n2, 239
Axelos, Kostas, 170

Bachelard, Gaston, 125, 167
Bacon, Roger, 117, 138
Banu, Georges, 165, 231n11, 232n49, 239, 250
Barrow, John D., 73, 223n27, 224n214, 239
Baruzi, Jean, 219n13, 239
Baudrillard, Jean, 195
Bataille, Georges, 169
Beckett, Samuel, 16, 158–59, 165
Bell, J. S., 42, 57, 223n14, 223n16, 239
Benrath, Frédéric, vii, 174–75, 234n42, 246
Berdiaev, Nikolai, 30, 219n32, 220n33, 239, 240

Berger, René, 35
Birkhoff, George David, 128
Blanchot, Maurice, 169
Boehme, Jacob, 6, 22, 24, 28–30, 33, 47, 97, 148, 209, 217n3, 219n3, 219n8, 219n10, 219n20–29, 219n31–32, 220n33–34, 220n43–44, 239–40, 248
Bohm, David, 152
Bohr, Niels, viii, 33, 41, 48, 54, 125, 140, 142, 152, 184–86, 193, 226n1, 235n20, 240
Boole, George, 77
Born, Max, 27, 41, 51, 219n18
Bosquet, Alain, 170, 232n16, 233n19, 240
Brancusi, Constantin, 16, 100
Brenner, Joseph E., 236n15, 237n30, 240, 245
Breton, André, vii, 125, 167–70, 232n1, 232n3, 232n5–6, 232n9–10, 240, 246, 249
Bricmont, Jean, 195–96, 235n7, 235n9, 251
Brody, T. A., 227n9, 240
Brook, Peter, vii, 12, 153, 155–66, 218n27, 230n1–2, 230n4, 231n5, 231n7, 231n20–21, 231n28, 231n35–40, 232n49, 239, 244, 246, 248, 250
  The Empty Space, 231n5, 231n8–11, 231n13, 231n16, 231n18, 231n22, 231n31, 232n44–45, 232n47–48, 232n50–51, 240
Bruno, Giordano, 72

Caillois, Roger, 169

NAME INDEX

Calder, Nigel, 95, 224n18, 240
Camus, Michel, 6, 13, 217n15, 218n28, 228n28, 240
Cardano, Gerolamo, 7, 84–85, 218n19, 241
Carroll, Lewis, 76, 150
Carter, Brandon, 93
Cassirer, Ernst, 206
CERN: European Center of Nuclear Research, 56, 60, 224n15, 244
Chekhov, 158
Chew, Geoffrey, 33, 88–91, 95, 220n45, 224n4–5, 224n7–10, 224n20, 228n31, 240–41, 243, 244, 245
Cilliers, Paul, 2, 225n1, 236n13, 241
CIRET: Centre International de Recherches et Etudes Transdisciplinaires, 220n48, 237n30, 241, 245
CIRT: Centre International de Recherches Théâtrales, 160, 162
Clair, Jean, 80–81, 223n10, 241
Connes, Alain, 71, 222n20, 241
Conrad, Joseph, 81
Copernicus, Nicolaus, 5, 138
Corbin, Henry, 11, 187, 218n25, 235n24–25, 241, 242
Crafton, Anthony, 218n19, 241
Cushing, James T., 95, 224n6, 224n19, 241

d'Ambrosio, Ubiratan, 35
Dali, Salvador, vii, 79, 125, 170–75, 233n19–20, 233n22–24, 233n26–29, 233n33, 233n35–36, 234n37–41, 240, 242–43, 246, 250
 50 Secrets of Magic Craftsmanship, 172–73, 233n35, 242
 Anti-Matter Manifesto, 171, 233, 242
 and nuclear mysticism, vii, 170, 172, 233, 242, 245
 Corpus Hypercubus, 172–73
 Leda Atomica, 172
 's theological writings, 171
Dallaporta, Nicolo, 35

Damasio, Antonio R., 160, 228n35, 231n33, 242
Daumal, René, 186, 235n23, 242
Davies, Paul, 110, 122n6, 225n14, 242
da Vinci, Leonardo, 173–74, 233n33, 246
de Brabant, Siger, 169, 232n13
de Broglie, Louis, 41, 50, 218n2, 232n16, 243
de Gaigneron, Ludovic, 26, 219n16, 242
de Nerval, Gérard, 138
Deledalle, Gérard, 117, 225n5, 226n22, 249
Deleuze, Gilles, 193, 195
Democritus, 90
Derrida, Jacques, 193
Descombes, Vincent, 220n3, 242
d'Espagnat, Bernard, 54, 57, 142, 221n13, 221n15, 221n17, 242
Devi, Maitreyi, 35
Devos, Raymond, 124
Diaz, Pro, 175
Dicke, Robert H., 92–93, 224n12–13, 242
Dirac, P. A., M. 41, 89, 224n12–13, 242
Dobbs, B. J. T., 218n21, 242
Dostoevsky, Fyodor, 80
Duchamp, Marcel, 81–82
Durand, Gilbert, 30, 35, 125, 130, 182, 187, 220n36, 227n12, 235n11, 235n25, 242
Dürrenmatt, Friedrich, 23, 219n7, 243
Dyson, Freeman, 199

Eco, Umberto, 138, 144, 228n1, 229n22, 243
Eddington, Arthur Stanley, vii, 87–88, 152, 223n1–2, 224n3, 243, 251
Einstein, Albert, 68, 75–77, 79–83, 85–86, 126, 141, 156, 171, 178, 181–82, 184–85, 190, 195, 206, 218n2, 219n6–7, 219n15, 219n18, 220n1, 220n39, 243, 245

NAME INDEX

Eisele, C., 243
Eliade, Mircea, 14–15, 218n29, 218n31–32, 243, 250
Eliot, Thomas Stearns, 169
Engels, Friedrich, 39

Feyerabend, Paul Karl, 193
Fichte, Johann Gottlieb, 6
Fierz, Markus, 140–42, 228n4, 229n5–6, 229n11, 229n14
Finkelstein, David, 52, 221n12, 243
Fischer, Carol Anne, 230n26, 243
Fodor, Jerry, 196
Freud, Sigmund, 171

Gablik, Suzi, 11, 218n26, 236n15, 243
Gadamer, Hans-Georg, 201, 205–6, 236n20, 243
Gale, George, 96, 224n20, 243
Galilei, Galileo, 3–6, 37–38, 41, 75, 142
  and the methodology of modern science, 5
  Dialogue on the Two Systems of the World, 3, 37, 217n1, 220n1, 243
Gell-Mann, Murray, 65, 68, 199, 222n12, 222n14, 225n5, 243
Georgi, Howard, 61
Ghyka, Matila, 156, 172–73, 230n3, 233n30, 233n32, 243
  The Geometry of Art and Life, 172, 233n31, 243
Glashow, Sheldon L., 61, 65, 70, 222n11, 222n17, 244
Gleizes, Albert, 81
Gödel, Kurt, 223n28, 248
  's theorems, 73, 106
Goethe, Johann Wolfgang von, 6, 138
Goldman, Steven Louis, 5, 217n13, 244
Gospel of Thomas, The, 189, 192, 235n3, 250
Granger, Gilles Gaston, 132, 228n30, 244
Greene, Brian, 70, 73, 222n15–16, 223n30, 244

Greene, Michael, 68
Gross, David, 133, 228n31, 244
Guattari, Felix, 193, 195
Guénon, René, 30, 34, 191, 220n35, 220n46, 235n1, 244
Gurdjieff, George Ivanovich, 155
Guth, Alan H., 109, 225n13, 244

Hadamard, Jacques, viii, 10, 84, 178, 180–82
  An Essay on the Psychology of Invention in the Mathematical Field, 84, 178–79, 218n24, 234n2, 234n4–9, 244
Hantaï, Simon, 169
Hartmann, Nicolai, 205, 236n3, 244
Hawking, Stephen, 94, 172, 224n15, 244
Heidegger, Martin, 138, 150, 169, 205, 206
Heilpern, John, 160, 231n7, 231n17, 231n29, 231n34, 244
Heisenberg, Werner, 89, 106, 129, 171, 193, 205–7, 244
  Manuscript of 1942, 106, 225n11, 236n4, 244
Henderson, Linda Dalrymple, 81, 171, 223n11, 233n21, 244
Henry, Michel, 191, 235n2, 244
Heraclitus, 167, 170
Higgs, Peter, 60–61
Hinton, Charles Howard, 77–80, 82–83
  Scientific Romances, 223n6, 244
  The Fourth Dimension, 77, 223n5, 244
Höffding, Harald, 185
Hoffmeyer, Jesper, 214, 237n32, 244
Holton, Gerald, 32–33, 185, 219n18, 220n39–42, 235n16, 235n18–19, 235n21, 245–46
Horgan, John, 198–99, 236n14, 245
Horia, Vintila, 129, 227n11, 227n20, 241, 245
Husserl, Edmund, 106, 122, 206, 208, 236n14, 245

Huyghe, René, vii, 125, 174–75, 234n47–48, 245

Ionesco, Eugène, vii, 147–50, 154, 229n4–8, 230n9, 230n23, 234n36, 245, 247
Irigaray, Luce, 195

Jacquart, Emmanuel, 150, 230n9, 245
James, William, 185, 235n17, 235n19, 245
Jammer, Max, 185, 221n7, 245
Janés, Clara, 2, 213, 237n29–30, 245
Janet, Pierre, 185
Jantsch, Erich, 104, 225n7, 245
Jaspers, Karl, 169
Jean-Blain, Marguerite, 148, 229n4, 245
Saint John Climacus, 205
Saint John of the Cross, 22, 25–26, 148, 219n13, 239, 245
Pope John Paul II, 196
Jouffroy, Alain, 175
Joyce, James, 118
Juarroz, Roberto, 147, 229n1, 245
Jung, Carl Gustav, 131, 138–44, 152, 160, 169, 229n7–8, 229n16–19, 245–46

Kaku, Michio, 223n1, 245
Kaluza, Theodor, 62
Kane, Sarah, 154
Kepler, Johannes, viii, 3, 7, 38, 138, 182–84, 186, 218, 220n39, 235n13, 235n15, 245, 250
    and the living Earth, 182
    The Cosmographical Mystery, 184
    The Harmony of the World, 184, 235n15, 245
Khun, Thomas, 185, 193
Kierkegaard, Søren, 185
Klein, Oscar, 62
Knight, Christopher C., 212, 237n17, 246
Kojève, Alexandre, 4–6, 217n4, 217n13, 244, 246

Korzybski, Alfred, 31, 125, 220n38, 226n2, 246
Kremer-Marietti, Angèle, 32, 220n41, 246
Kristeva, Julia, 195
Kupka, František, 81
Kustow, Michael, 218n27, 246

Lacan, Jacques, 193, 195–96
Lamartine, Alphonse de, 25
Laplace, Pierre-Simon, 53
Latour, Bruno, 193, 195
Laurikainen, K. V., 141–42, 229n5–6, 229n11, 229n13–14, 229n20, 246
Laszlo, Ervin, 225n2, 225n6, 246
Lederman, Leon, 61, 72, 222n4, 222n8, 222n25, 246
Lee, T. D., 132
Leibniz, Gottfried Wilhelm von, 38, 48, 95–97, 224n20, 243
Lenin, Vladimir Ilyich, 80, 196
    Materialism and Empiriocriticism, 196
Leupin, Alexandre, 5, 217n11, 246
Lissitzky, Lazar, 81
Lomas, David, 173, 233n33–34, 246
Lucretius, 102
Lupasco, Stéphane, vii, 6, 121–23, 125–30, 132, 139–40, 142, 151–52, 154, 167–71, 174–75, 193, 217n16, 219n30, 226n3, 226n27, 227n3–8, 227n10, 227n12–15, 227n17–18, 227n21–24, 228n25–28, 229n12, 229n15, 232n2, 232n4, 232n7–10, 232n15–16, 233n23–24, 233n36, 234n42–46, 234n49, 239–43, 246–48, 250
Lyotard, François, 193

Magnin, Thierry, 6, 125, 217n17, 228n28, 240, 246
Magritte, René, 169
Malevitch, Kasimir, 80–81, 273n10–11, 241, 244
Malherbe, Jean-Francois, 120, 135, 226n38, 227n22, 228n36, 232n14, 247

NAME INDEX

Mallarmé, Stéphane, 25
Malraux, André, 15, 169
Mansour, Joyce, 169
Marcel, Gabriel, 169
Marcus, Solomon, 225n1, 247
Marx, Karl, 6, 39
  Marxist, 4, 39, 123
Mathieu, Georges, vii, 125, 169–70, 175, 232n12–13, 232n15–16, 233n17–18, 244, 246–47
Matrix, The, 78
Matta, Roberto, 175
Mauclair, Jacques, 149–50
Metzinger, Jean, 81
Michaux, Henri, 125, 170, 175, 186, 235n22, 247
Minkowski, Hermann, 82–83, 85, 156, 223n12, 247
Modreanu, Simona, 154, 230n23, 247
Mondrian, Piet, 81
Morales, Gregorio, 151–53, 230n15, 230n18, 230n20–22, 247, 250
Moraru, Christian, 14, 212–13, 247
  Cosmodernism, 212, 218n30, 237n19–28, 247
Morin, Edgar, 102, 106, 125, 134, 225n3, 225n9, 227n23, 228n32–33, 247–48

Nagel, Ernest, 248
Needham, Joseph, 3, 217n2, 248
Ne'eman, Yuval, 95, 224n17, 248
Newman, James R., 223n28, 235n12, 248
Newton, Isaac, 3, 6–7, 38, 48, 63, 65, 89–90, 138, 183, 195, 218, 242
Nottale, Laurent, 65–66, 222n13, 249
Novalis, 6

Ouspensky, Pyotr Demianovitch, 80, 219n9, 220n37, 223n8, 223n10–11, 241, 244, 249
  The Fourth Dimension, 80, 244
  Tertium Organum, 80, 223n8, 249

Pagels, Heinz, 58, 221n19, 222n3, 249

Panikkar, Raimon, 17, 218n37, 249
Parinaud, André, 167, 170, 232n1, 240
Parkinson, Gavin, 175
  Surrealism, Art and Modern Science, 175, 234n50, 249
Parmenides, 170
Paulhan, Jean, 169
Pauli, Wolfgang, 1, 41, 126, 129, 131, 139–45, 200, 228n4, 229n5–6, 229n9–11, 229n14, 236n16, 239, 246, 249, 251
Peirce, Charles Sanders, vi, 14, 97, 113–19, 183, 214, 225n2–5, 226n7–13, 226n16–26, 226n28–29, 226n32–37, 237n34, 243, 249, 250–51
Penrose, Roger, 131, 221n20, 227n16, 249
Picasso, Pablo, 81–82
Planck, Max, v, 41, 45, 47–50, 62, 69, 85, 109, 126, 170, 190, 220n4, 221n1, 221n4, 221n10, 228n34, 249
Plato, 4
Poincaré, Henri, 64, 178, 180, 186, 234n1, 234n3, 249
Popper, Karl, 199
Priestley, John Boynton, 80, 223n9, 250
Prigogine, Ilya, 174, 199
Primas, Hans, 141, 229n10, 239
Proust, Marcel, 81
Puech, Henri-Charles, 235n3, 250

Ragon, Michel, 175
Ramanujan, Srinivasa, 182, 235n12, 248
Random, Michel, 221n1, 250
Régy, Claude, 154, 230n25, 250
Rennie, Bryan S., 218n32, 250
Resnais, Alain, 154
Riemann, Georg Bernhard, 68, 75–76, 81, 186
Riger-Brown, Allan, 153, 230n22, 250
Rothko, Mark, 81
Rubbia, Carlo, 60

Rûmî, Djalâl-ud-Dîn, 19, 218n1, 250
Russel, Bertrand, 16

Sakata, S., 95, 224n16, 246
Salam, Abdus, 35, 59–61, 92, 221n1, 222n3, 222n7, 224n11, 250
Sarraute, Nathalie, 147–48, 229n2, 250
Sartre, Jean-Paul, 168
Saussure, Ferdinand de, 118
Schelling, Friedrich Wilhelm Joseph von, 6, 138
Scherk, Joël, 68–69
Schiller, Johann Christoph Friedrich von, 186
Schlegel, Karl Wilhelm Friedrich von, 6
Schöffer, Nicolas, 175
Schrödinger, Erwin, 41, 47–49, 51, 89, 172
Schwarz, John, 68–69
Science and the Boundaries of Knowledge: The Prologue of Our Cultural Past, 35, 220n47–48, 241, 250
Scofield, Paul, 158
Scriabin, Alexander, 81
Segré, Emilio, 49, 221n9, 250
Serres, Michel, 193
Shakespeare, William, 158–59
Shlein, Leonard, 151, 230n16, 250
  Art and Physics: Parallel Visions in Space, Time, and Light, 230n16, 250
Serban, Andrei, 166, 232n52, 247
Silburn, Lilian, 48, 221n6, 250
Smolin, Lee, 72–73, 222n23, 229n26, 250
Snow, C. P., 8–9, 194, 218n22, 251
Sokal, Alan, 193–97, 235n4–5, 235n7–11, 251
  Affair, viii, 193–97
  's Hoax, 195, 235n6, 251
Soulages, Pierre, 175
Stapp, Henry, 35, 221n20, 251
Stearns, Isabel, 117–18, 226n23, 226n30, 251
Steiner, George, 16, 218n35, 251
Susskind, Leonard, 68, 222n21, 251

Sypher, Wylie, 150–51, 230n10, 251

Tàpies, Antoni, 175
Tegmark, Max, 221n18, 251
Teilhard de Chardin, Pierre, 99, 112, 225n16, 251
Thirring, Walter, 119, 226n31, 251
Thom, René, 172–74
Tipler, Frank J., 93, 112, 224n14, 225n17, 239, 251
Townsend, Paul, 71
Transdisciplinarity: Theory and Practice, 218, 240

van der Meer, Simon, 60
van Doesburg, Theo, 81
von Neumann, John, 128
Varese, Edgar, 81
Veneziano, Gabriele, 68, 92
Villon, Jacques, 81
Virilio, Paul, 193, 195
von Baader, Franz Xaver, 138
von Bertalanffy, Ludwig, 102, 225n4, 251
von Franz, Marie-Louise, 229n21, 251
von Weizsäcker, C. F., 49, 221n8, 251
Voss, Karen-Claire, 2, 6, 218n23, 240, 248

Weinberg, Steven, 60–61, 108, 110, 194–95, 197, 199, 222n2, 225n12, 225n15, 235n6, 236n12, 251
Wells, H. G., 76
Wheeler, John Archibald, 67, 221n18, 224n20, 243, 251
Whittaker, E., 223n2, 251
Wiener, Norbert, 232n16
Wilde, Oscar, 81
Witten, Edward, 70–71, 199, 222n18, 251
Wittgenstein, Ludwig, 135

Yeats, William Butler, 138

Zeami, 159, 162, 165, 231n27, 232n43, 232n46, 252

SUBJECT INDEX

actualization, 13, 34, 116, 118, 126, 128–29, 131–32, 134–35, 143
antagonism
  logic of, 126–27
anthropic principle, vi, 92–95
  as special case of the bootstrap principle, 94
axiom
  of identity, 121, 145
  of non-contradiction, 121, 124, 128, 144–45
  of the excluded third. See excluded third
  of the included third. See included third

Bell's inequalities, 57, 221n16, 239
bootstrap, vi, 33–34, 68, 88, 89, 90–92, 95–97, 102, 220n45, 224n8–9, 224n19, 240–41
  and Leibniz's monadology, 96
  and Taoism, 97
  and the anthropic principle. See anthropic principle
  cosmic, vi, 33, 107, 110
  hadron, 90, 92
  principle, vi, 33–34, 87–88, 90, 92, 94–97, 104, 116
  theory, 68, 89, 92, 95
  total, 97

causality
  deterministic, 184
  formal, 38
  global, 42, 121, 128, 131, 190
  local, 38–39, 41–43, 49, 57, 121, 128, 190

violation of, 57
civilization, civilizations
  atheistic, 16
  Chinese, 3, 12
  Islamic, 12
  war of, 16, 148
  Western, 141, 169
Coincidentia Oppositorum, vii, 138–39, 152
complementarity, viii, 33, 184–86, 228n28, 241
  Bohr's principle of, 33, 125, 140, 152, 167, 184, 185–86
  contradictory, 34
complex
  plurality, vi, 14, 99, 106
  reality. See reality
  structure, 207
  system, systems, 138, 191, 225n1, 241
  unity. See unity
  world. See world
complexity, 2, 7, 66, 99–104, 107, 112, 124, 130, 178–79, 192, 211
  and fragmentation, 30
  and postmodernism, 225n1, 241
  and Reality. See Reality
  and space-time, 104
  and transdisciplinarity. See transdisciplinarity
  as a modern form of the principle of universal interdependence, 211
  degrees of, 107
  experimental, 100
  in science, 101
  in the natural systems, 102
  levels of. See levels

complexity (continued)
  logic of, 124
  mathematical, 100
  multidimensional, 101
  of the universe, 101, 153
  organized, 104
  philosophy, 225n1
  simplicity/, 32, 209
  simplicity and, 101
  social, 101
  types of, 104, 107
consciousness
  consciousness of, 168, 175
  cosmodern. See cosmodern
  flow of, 14, 208
  global, 185
  hysteric's, 186
  history of, 15
  immediate, 117
  levels of. See levels
  matter-, 209
  of being home, 207
  planetary, 2
  science of, 221n20, 249
  source of, 15
  structure of, 15
  transpersonal, 2
  visionary, 2
contradiction, contradictions, v, 20, 23–24, 27, 34, 47, 70, 75, 121, 126, 128, 132, 135, 144, 147, 149, 151, 158–59, 168, 174, 193–95, 197, 210, 227n8, 227n21, 227n23, 228n25, 241, 243, 246–47
  actualization-potentialization, 132
  at the origin of the world, 209
  between art and science, 156
  between Three and Four, 144
  internal, 73, 90
  logical, 168
  logic and, 148–50, 168, 232n4, 232n7
  logic of, vii, 123, 147, 151, 154, 167–68, 175
  non-, 121, 123–24, 128, 144–45, 149–50, 167, 174, 191, 206
  wave-corpuscle, 33
  unity of. See unity
  unity in, 155
  universal, 139
contradictional
  conjunction, 123
  disjunction, 123
contradictories, 67, 121–22, 124, 130–31, 143–44, 185, 208–9
  mutually exclusive, 121
  unification of. See unification
  supreme, 209
  unification of. See unification
  unity of. See unity
Copenhagen interpretation, 54
cosmodern, 212
  consciousness, 215
  culture, 10, 14
  era, 212
  imaginary, 212
  inhabitant of the world, 16
  mind, 212
  rationality, 213
  view of Reality. See Reality
  vision, 14
cosmodernism, 212–13, 218n30, 237n19, 247
cosmodernity, viii, 190, 192, 198, 200, 203, 209, 210 fig. 17.4, 212–14
  ethical mperative of, 213
  foundation of, 212
cosmos, 43, 212, 214, 224n14, 239
  Cardano's, 218n18, 241
  Chinese vision of the, 3
  evolution of. See evolution
  metaphysical, mythological, and metaphorical idea of, 37
  objective order in the, 140
  particle and, 62
culture, cultures
  abyss between, 11
  antagonistic, 9–10
  and philosophy, 195
  and religions, 201
  and spirituality, 6
  cosmodern. See cosmodern

dialogue between, 12
dialogue between science, culture, spirituality, and society, 107
humanistic, 8–10
intermingling of, 11
monolithic, 8
new vision of, 106
of hope, 12
postmodern, 9
quantum. *See* quantum
science and, v, 7–8, 11, 154, 173, 194–95, 197
scientific, 8–10
shattered, v, 3
spiritual, 8–9, 236n19, 249
technoscientific, 8–9, 201
  conversion to values, 201
the two, 8–9, 218n22, 251
traditional, 11
trans. *See* transculture
translation between, 11
war of, 16
world of. *See* world

democracy, 200
  dimensional, 71
  nuclear, vi, 71, 91–92
  spiritual dimension of, viii, 199–200
determinism, 38–39, 42–43, 49, 53
  absolute, 53
  and freedom, 173
  classic, vi, 52, 89
  mechanistic, 156, 190
discontinuity, v, 26, 28, 41, 47–48, 54, 67, 126, 131, 163, 190, 215, 236n13, 241
  and continuity, 27, 50, 121, 164
  continuity/, 34
  in Indian philosophical thinking, 48
  of the physical laws, 26
  principle of, 28, 29
  quantum. *See* quantum
  real, 41
  that announces a change in history, 126

diversity
  indefinite, 30
  identity and, 132
  through unity, 20, 103
  unity and. *See* unity
  unity in. *See* unity

evolution, 99, 103, 142, 164, 228n28, 241
  according to Peirce, 119
  biological, 80
  historical, 126
  in theater, 159
  /involution, 32
  and involution, vi, 110, 112, 144
  of cosmos, 108
  of physics, 46, 49
  of rationality, 26
  of science, 6
  of scientific ideas, 32
  of natural systems, 42, 134
  of the universe, 92, 94
  spiritual, 20

field theory
  of gravity, 64
  quantum, 96, 102
Firstness, 115, 116–19, 226n23, 226n30, 251
  as language of Tradition, 120
fourth dimension, vi, 62, 75–81, 86, 152, 223n5, 223n11
  in modern art, 171, 233n21, 244
fragmentation
  accelerated, 26, 190
  cultural, 11
  disciplinary, 11, 99
  of knowledge, 214
  of the human being, 214

God, 39, 51, 59, 61, 77, 152–53, 190, 192, 198, 207, 209, 225n17
  almighty, 191
  death of, 200
  emanation of, 144
  existence of, 80

God *(continued)*
  human/, 40
  hypothesis, 190
  messangers of, 205
  metaphors, 206
  multidimensional, 78
  of nature, 212, 237n17, 246
  particle, 61, 222n4, 222n8, 222n25, 246, 251
  religion without, 195
  's being, v, 24
  's body, 24
  's death, 198
  's existence, 207
  's secret, 27
  's uniqueness, 4
  's wisdom, 29
  unique, 102
Gödelian structure of nature and knowledge, 107, 116, 223n29, 248

Heisenberg's uncertainty relations, 31, 67, 87, 105
heterogeneity, 117–18, 127, 129, 134–35
Hidden Third, viii, 2, 170, 201–3, 208, 209–15, 210 fig. 17.4, 211 fig. 17.5, 236n13, 237n30, 241, 245
Homo
  economicus, 200–1
  religious, 200–1
homogeneity, 117–18, 128–29, 134–35

imaginal, viii, 186–87, 235n24, 235n25, 241–42
imaginary, viii, 3, 78, 141, 143, 152–53, 177–79, 182, 185–87, 200
  cosmodern. *See* cosmodern
  in artistic creation, 179, 186
  in scientific creation, 180–81, 186
  new cultural, 218n30, 247
  numbers, 7, 84–85
  real/, 202
  Reality of the. *See* Reality
  spacelike distance, 85
  time, 86

imaginatio vera, 202
imagination, 8–9, 40–41, 77, 103
  as a fold in the real, 202
  Dali's, 172
  image and, 177
  ordinary, 41, 48
  scientific, 219n18, 220n39, 220n40, 235n16, 235n18–19, 235n21, 245
  symbolic, 220n36, 227n12, 242
  visionary, 202
incompleteness, 208, 210
infinitely
  big, 62–63
  brief, 1, 41
  large, 1, 27, 101
  long, 1
  small, 1, 27, 41, 46, 53, 62–63, 101
interconnectedness
  continous, 211
  of Reality. *See* Reality
  universal, 27
  universe of, 192
intrareligious dialogue, 17, 218n37, 249
irrational, vii, 23–26, 39, 57, 119, 132–34, 143, 165, 189, 197
  as an appeal, 132
  as an obstacle, 132
  asymptotic, 133
  by cancellation, 132
  cooperation between rational and, 133
  force 190, 200
  of globalization, 8
  types of, 132
irrationalism
  hermetic, vii, 138–39
irrationality
  kernel of, 134
  in scientific knowledge, 49, 134
irrationalizable, 134
isomorphism, viii, 34, 130–31, 185, 197
  between different planes of knowledge, 191
  between the different sciences, 102, 192

between the quantum physics and
  art, 172
between the quantum world and the
  world of the psyche, 142
between the supersensible world and
  the world of events, 198
general, 30
laws of, 30

johakyu, 159, 162–64

language, 118–19
  accuracy of, 120
  and Reality. See Reality
  artificial, 6, 38, 114
  as a quantum phenomenon, 118
  between God and humans, 6, 38
  Boehme's, 6
  classical, 46
  entropy of, 31
  inexact, 119
  macroscopic, 46, 54
  mathematical, 41, 113–14, 119–20,
    151, 193, 211
  modern, 29
  natural, vi, 21, 30, 113–14
  of physics, 60
  of science, 120, 194
  of Thirdeness. See Thirdeness
  of Tradition, 120
  ordinary, 22, 31, 47
  realism of, 5
  scientific, vi, 22, 31, 113
  structures, 125
  symbolic, 119, 211
  theory of, 114
  trans. See translanguage
  universal, vii, 12–13, 118–20
  verbal, 150–51
  Wittgenstein's, 135
  wooden, 199
level, levels
  emotional, 205
  inorganic, 205
  intellectual, 205
  of being, viii, 205

of complexity, 107, 138
of consciousness, 159
of experience, 157
of knowledge, 191
of materiality, 31, 51, 198
of organization, 107, 206
of our luminous ignorance, 13
of perception, 13, 164, 208
of Reality. See Reality
of representation, 9–10, 217n15,
  228n28, 240
of silence, 13
of spontaneity, 164
of understanding, 139, 215
organic, 205
quantum. See quantum

matter
  classical concept, 190
  contemporary concept, 103, 190
meaning
  as dynamic source of Reality, 166
  exact, 114
  horizontal, 201
  inaccuracy of, 117
  inside narrow dogmatic frameworks, 31
  meaning of, 201
  of history, 15
  of science, 2
  of life, 2, 7, 189, 200
  of theater, 155
  of wars, 148
  science and, 7
  vertical, 201
methodology, 11, 37, 96, 212
  apophatic, 237n31, 248
  deviant, 96
  hypothetical-deductive, 96
  of modern science, 4–5, 7
  systemic, 104
  transdisciplinary. See transdisciplinary
modernity, viii, 1, 3, 7, 12, 16, 37, 142,
    200, 203, 204 fig. 17.2, 212
  cos. See cosmodernity
  post, viii, 203, 204 fig. 17.3, 212
  pre, viii, 203, 204 fig. 17.1

movement
  and discontinuity, 5, 26, 28
  and interrelations, 155
  as a perpetual genesis, 30
  harmonious, 163
  horizontal, 163
  of information, 116
  of information and consciousness, 15
  ordered, 214
  perpetual, 118, 197
  quantum, 171
multiverse, 72
mystery, 71, 169
  epiphany of a, 30
  irreducible, 197, 215
  of our consciousness, 80
  of the present moment, 158
  of the universe, 95
  theorists, vi, 70
  unfathomable, 48

new barbarity, viii, 190–91
nonlocal correlations, 42, 57
nonresistance, 106, 201, 203, 205, 207–10, 212
nonseparability. See quantum

Object, viii, 14, 197, 201–3, 204 fig. 17.1–17.3, 206, 208–9, 210 fig. 17.4, 210, 212, 215
  levels of Reality of the, 9, 201, 207–8, 215
  Subject/. See Subject
  transdisciplinary, 208, 210
Otherness, 215

potentialization, 34, 118, 126, 128–29, 131–32, 135
principle of relativity, 106
principle of universal interdependence. See complexity
psychophysical problem, the, vii, 137, 139, 141, 143

quanton, quantons, 42–43, 123

quantum, 41, 48, 54, 221n18, 251
  aesthetics, vii, 147, 151–52, 154, 230n20, 247
  artist, 152
  chromodynamics, 96
  cosmology, 101, 108–9
  culture, 152, 230n17, 230n18, 240, 247
  dialectic, 129
  discontinuity, 172
  effect, effects, 53, 58
  entities, 27, 42–43, 89, 105
  event, events, 13, 27, 43, 51–53, 105, 111
  field theory. See field theory
  fluctuations, 55, 67, 69, 104
  foam, 67
  freedom, 52, 111
  geometry, 68
  gravity, 63–65, 193, 235n4, 251
  indetermination, 131
  indeterminism, 104
  infraworld. See world
  interactions, 42
  laws, 24, 46, 58, 105
  leap, 49
  level, 41, 43, 186
  level of materiality. See level
  logic, 122, 128, 221n12, 227n9, 240, 243
  mechanics, 1, 25, 27, 40–41, 43, 46, 48, 50–51, 53–54, 57, 63–69, 71–72, 89, 116, 121–22, 128, 139–41, 153–54, 171–72, 175, 194–95, 206, 221n7, 221n14, 221n20, 222n18, 234n50, 239, 245, 249, 251
  nonseparability, vi, 42, 56–58
  numbers, 67
  of action, 26, 48–50, 52
  paradox, paradoxes, 6, 47, 129, 185
    Einstein-Podolsky-Rosen, 47
    Schrödinger's cat, 47
  particle, particles, 43, 50–53, 55–56, 172, 184

phenomena, 27, 51–52, 54, 184–86
philosophy, 221n14, 239
physics, vi, 6, 8, 41, 51, 54–55, 57, 62, 64, 66, 100–2, 104, 106, 115, 121–22, 126, 128–30, 172, 185, 190, 197–98, 201, 215
Planck's, 48
principle of superposition, 58
processes, 58, 108
randomness, 43
realism, 47
representation, 51
revolution. See revolution
scale, 27, 31, 55, 104, 108, 110–11
spontaneity, v, 50, 52, 58
state, 129
superposition, 51
surrealism. See surrealism
system, systems, 89, 118
theater. See theater
theory, 27–28, 66, 96, 158, 167, 221n8, 221n12, 221n15, 243, 251
thinking. See thinking
universe, 190, 192
vacuum, vi, 13, 54, 56, 104, 109, 172
values, vi, 53
view of the world. See world
vision of the world. See world
world. See world

Real, 105, 203, 207
and Reality. See Reality
reality, Reality, vii, viii, 1, 12, 14, 19, 21–22, 25–26, 31–32, 37, 40, 42, 48, 54, 56–57, 62, 69, 75, 78, 85, 87–88, 95–96, 102, 105, 107, 112, 114, 118–20, 126, 128–31, 133, 140–43, 145, 151, 153, 155, 158, 161, 163, 165, 168–69, 174, 179, 181, 183, 185, 187, 194–95, 201–3, 205–9, 211, 214–15, 221n13, 221n15, 236n1, 236n2, 236n15, 240, 242
abstraction is a component of, 32
complex, 99
complexity and, vi, 99
contemporary view of, 205
contradictory aspects of, 34
cosmodern view of, 214
direct perception of, 21, 151
dissolution of, 32
dynamical source of, 116
eternal genesis of, v, 26
existence of, 114
ever-changing, 192
gradual revelation of, 32
hidden, 152
hidden face of, 19
human, 34
image of, 119
 classical, 198
incommunicable, 22
independent, 82
 of man, 57
indivisible, 23
inseparable, 57
interconnectedness of, 211
is plastic, 215
Jacob Boehme's view of, 29
language and, 114
level of, levels of, vi, 9, 10, 13–15, 25–27, 31, 40, 43, 51, 100, 104, 105, 106–7, 119, 121, 123, 128, 137–38, 143–45, 153–54, 158, 196, 200–2, 205–12, 214–15, 217n15 236n13, 236n16, 240–41, 248–49
and the Hidden Third. See Hidden Third
 cosmic, 214
 discontinuous, 211
 individual, 214
 in the Heisenberg model, 206
 of the Object. See Object
 of the Subject. See Subject
 spiritual, 214
 transdisciplinary theory of, 201, 214
 unified theory of, viii, 207, 214

266   SUBJECT INDEX

reality, Reality *(continued)*
    mathematical, 62
        structure of, 186
    multidimensional, 38
    multiple, 215
    multireferential structure of, 207
    nature of, 33
    new model of, 140, 142
    objective, 40, 195
    of physical laws, 194
    of the imaginary, 186
    of the symbol, 140
    of vision, viii, 177
    one-dimensional, 100
    ordered movement of, 214
    perception of, 181
    physical, 34, 53, 69, 89, 143
    psyche, 143
    rational, 215
    Real and, 105, 203
    regions of, 106, 206
    scale of, 205
    separate, 21
    social construction of, 56
    source of, 215
    spiritual, 165
    subjective, 206
    systemic approaches to, 103
    ternary dialectics of, vii, 128
    ternary structure of, 115–17, 124
    theatrical, 155–56
    three-dimensional, 62
    trans-, 210, 211 fig. 7.5, 212
    transdisciplinary, 210, 214
        approach to, 201, 209
    truth and, 115, 195–97
    unique, 21, 23, 87
    view of, 1, 21
    vision of, viii, 177, 203
reduction, vii, 137, 139, 211
    chain, 137
    non-, 236n16, 248
    scientific meaning, 137
    wave packet, 54
reductionism, vii, 103, 137–38, 211
    anti, 138

    inter, 137
    holism/, 32, 209
    multi, 137
    non, 211
    philosophical, 137
    scientific, 137
    trans. *See* transreducionsim
reductionist, reductionists 190
    currents, 24
    multi, 211
    thinking. *See* thinking
Renaissance, 3, 7, 62, 84, 101, 169, 218n19, 241
    New, viii, 215
revolution, 196
    biological, 1
    conceptual, 54
    information, 1
    quantum, v, 1, 27, 40, 43, 47
    Russian, 40, 80
    scientific, 218n22, 222n15, 251
    social, 39
    string, 73
    superstring, 68, 70

sacred, 14–16, 28, 106, 133, 165, 200
    abolition of the, 15
    experience of the, 16
    presence of the, 14
    roots of the theater, 166
science, sciences
    and art, 9, 174, 187
    and culture. *See* culture
    and meaning. *See* meaning
    and religion, 49, 107, 132, 191, 192, 197, 201, 228n28, 232n14, 236n16, 236n18, 237n31, 241, 247–49
    and spirituality, 197
    and technology, 8
    and the boundaries of knowledge, 219n5, 220n47–48, 241, 248, 250
    and theology, 217n15, 240
    and tradition, v, 20, 21–23, 34, 35, 155
    and wisdom, 9

classical, 1, 32
end of, viii, 198–200, 236n14, 245
exact, 7, 194
fundamental, 8, 189
  hard, 7, 139, 197
  human, 7, 8, 223n29, 248
  modern, v, 3, 4, 5, 6, 7, 11, 27, 37, 63, 107, 108, 153, 175, 189, 196, 198, 217n13, 244
  birth of,, 5, 6, 7, 194
  Christian Origin of, v, 3, 4, 6
  methodology of, 7
  Surrealism, Art and Modern Science, 175, 234n29, 249
natural, 185
soft, 7, 100, 139, 200
subjective objectivity of, 193
super, 201
scientism, v, 2, 37, 39–40, 193–94, 196
Secondness, 115, 116–19, 226n23, 226n30, 251
  as category of existence, 117
  as language of science, 120
self-
  birth, 2, 40
  confrontation, 161
  consistency, vi, 88–91, 94–96, 108, 111, 199
    of the universe, 92, 95, 104
  creation, 110
  creativity, 103
  destructing, 197
  destruction, 2, 12, 40, 190
  destructive, 123
  initiation, 22, 161
  organization, 103–4, 111
    of the universe, 110
  organizing
    universe, 225n7, 245
  perception, 212
  protection of the secret, 27
space-time, 21, 57, 63–64, 66, 77, 81–82, 104, 109, 116, 132, 227n16, 249
  and matter, 190
  classical, 64

continuous, 104, 131
continuum, 82, 89, 102
cyber-, 105
dimensions of, 67, 71
discontinuous, 104, 154
distance in, 83–85
fractal, 66, 222n13, 249
genetic code of, 71
geometry of, 63
home of, 77
macroscopic, 91
motion in, 89
multidimensional, 100, 109, 196
nature of, vii, 67, 104, 131
structure, 64
precise localization in, 52. See also Heisenberg's uncertainty relations
spirituality, 2, 9, 40, 107, 137, 200, 214
  art and 20
  culture and. See culture
  forgotten, 191
  materiality and, 137
  metaphysical, 172
  new, 199–200, 215
  science and. See science
  theater of. See theater
Subject, viii, 13–14, 192, 197, 201–3, 204 fig. 17.1–3, 206, 208–10, 210 fig. 17.4, 212, 215
  levels of Reality of the, 9, 201–2, 207–8, 215
  /Object relation, 203, 209
  resurrection of the, 197
  transdisciplinary, 208, 210
  unification of the, 212
  unification of the transdisciplinary Subject and the transdisciplinary Object, 208
  unity of levels of levels of Reality of the Subject, 208
superstrings, vi, 62, 69–70, 75, 92, 95, 104–5, 109, 152, 244
  revolution. See revolution
  world of. See world
subrealism, 171, 246

surrealism, 167–68, 171
  and modern physics, 8, 175, 234n50, 249
  quantum, 154
symbol, symbols, 30, 31–34, 97, 113, 119, 130, 140, 143–44, 186, 200, 206, 218n29, 243
  and Kepler, 183
  and the bootstrap principle, 33, 97
  and thêmata, 32
  as epiphany of a mystery. See mystery, 30
  facets of, 33
  gnostic, 110
  -idea, 34
  indefinite number of aspects of a, 31
  infinite richness of, 186
  mathematical, 180–81
  mutilation of the, 31
  reality of the, 140
  world of. See world
systemogenesis, vii, 130
  unus mundus, 141, 209

technoscience, 1–2, 9, 11, 16, 100, 142, 191, 200–1
theater, vii, 147–50; 155, 159–61, 165, 215
  American, 232n52, 247
  anti, vii, 8, 147, 150
  contemporary, 230n26, 243
  deadly, 60
  holy, 156
  Ionesco's, 150, 154
  non-aristotelian, vii, 148–49
  of Becket, 158
  of Cruelty, 165
  of Peter Brook, vii, 155, 158–59
  of spirituality, 166
  of the absurd, 150, 230n23, 247
  of the world. See world
  quantum, vii, 151, 153–54, 230n26, 242, 243
  rough, 156
thêma, thêmata, v, 30, 32–33, 220n40, 220n42

thinking
  analytical, 102–3
  apophatic, 214
  Aristotle's, 169
  as infomation, 117
  atomistic, 101–2
  beyond, 151
  binary, 191, 211, 214
  Bohr's, 185
  classical, 30, 49
  Eddington's, 87
  essence of, 119
  event of, 116
  hermetic, 138, 144–45
  informational, 117
  intuitive, 207
  Jung's, 140
  Kepler's, 184
  life of, 118
  living and, 118
  logical, 180, 182
  Lupasco's, 127, 139
  mathematical, 113
  mythical, 120
  one-dimensional, 140
  open, 134
  Pauli's, 140
  philosophical, 48, 126, 129, 141, 205
  quantum, 197
  rational, 158
  reductionistic, 190
  Riemann's, 75
  scientific, v, 26, 30, 32, 90, 95
  symbolic, v, 30–32, 120, 130
  systemic, vi, 89, 97, 101–4, 206
  traditional, 21, 26, 30, 48
  Western, 167, 236n16, 249
  without words, viii, 180
  wordless, 181
  world of. See world
third, 6, 29, 113, 115, 117, 148, 170
  excluded, 121, 123–24, 145, 191
    axiom of the, 121, 124, 128
  experienced, vii, 134
  hidden. See Hidden Third

included, vii, 30–31, 106, 122–25,
  127–28, 132, 135, 148, 170, 193,
  208–9
  and Wittgenstein's language games,
    135
  axiom of the, 123, 130, 145
  philosophy of the, 135
  logic of the, 124, 139, 152, 210
  matter, 169
  in the humankind-nature relation-
    ship, 23
  universe, 115
Thirdness, 115, 117–19, 226n23,
  226n30, 251
  invariance and, vii, 116
  language of, 120
time
  absolute zero of, 108
  imaginary. *See* imaginary
  living, 14
  machine, 76–77, 223n3, 248
  non-, 14–15
tradition, traditions, v, 19, 20–24,
  33–35, 90, 120, 126, 127, 155,
  156, 159, 193, 213, 231n27,
  232n43, 232n46, 252
  Eastern, 24
  Newtonian, 91
  objective subjectivity of, 193
  Occidental, 47
  Oriental, 47
  origin of, 20
  religious, 16
  science and. *See* science
  Western, v, 19, 47
transcosmic unity. *See* unity
transcultural, v, 10, 12, 13–14, 16, 201,
  212
transculture, v, 3, 218n33, 240
transdisciplinarity, 2, 9–10, 16, 200–1,
  214, 218, 220, 232n14, 236n13,
  236n16, 236n18, 237, 240, 241,
  245, 247–49
  as transgression of duality, 209
  complexity and, 236n13, 241
  logical axiom of, 207

epistemological axiom of, 207
ontological axiom of, 207
Manifesto of, 218n23, 236n12,
  236n17, 248
transdisciplinary, 211–12, 214, 218n33,
  228n28, 237n33, 240–41, 252
  approach to Reality. *See* Reality
  hermeneutics, 201
  methodology, 201, 207
  Object. *See* Object
  quest, 197
  Reality. *See* Reality
  research, 201
  Subject. *See* Subject
  theory of levels of Reality. *See*
    Reality
transhistory, 14
translanguage, 13
transperception, 10
transpresence, 16
transrational, 215
trans-Reality. *See* Reality
transreductionism, 211
transreligious, 16, 213
  attitude, v, 14, 16
  vision of the world, 16
transsubjective, 105
T state, 123, 128–29, 131–32, 134–35,
  144

unification, 26, 49, 60–65, 71, 73,
  212
  between space-time and matter, 63
  grand, 65, 79
  of all physical interactions, 26, 62,
    68, 92, 108
  of contradictories, 17
  of different fields of knowledge, 125
  of electromagnetic interaction and
    gravitational interaction, 62
  of general relativity and quantum
    mechanics, 63–64
  of gravity, 65
  of heaven and earth, vi, 63
  of opposites, 143
  of physical interactions, 63

unification *(continued)*
  of quantum physics and gravitational theory, 62
  of strong, weak, and electromagnetic interactions, 61
  of the Subject. *See* Subject
  of the transdisciplinary Subject and the transdisciplinary Object. *See* Subject
  of the world. *See* world
  of two mutually exclusive contradictories, 142
  of weak and electromagnetic interactions, 60
  scale, 61
  theories of, 60, 62, 68, 73
unity, 30, 33, 43, 49, 58, 97, 103, 108, 111, 131, 143, 183, 209
  and diversity, 3
  and self-consistency, vi, 88
  between the quantum scale and the cosmological scale, 108
  between thought, body, and feelings, 160
  between time and non-time, causal and acausal, 15
  complex, 201
  cosmic, 102
  /hierarchical structure, 32–33
  in diversity, 20, 103, 164, 209
  of all existing levels of Reality, 158
  of being, 153
  of contradictions, 50, 97
  of contradictories, 27, 29, 33, 97, 154, 186, 197
  of knowledge, 1, 172, 201
  of laws, 42
  of levels of Reality of the Object. *See* Object
  of levels of levels of Reality of the Subject. *See* Subject
  of natural systems, 111
  of opposites, 185
  of the continuous-discontinuous pairing of contradictories, 67
  of the whole universe, 58
  of the world. *See* world

  open, 10, 14–15, 105–6, 201, 209
  plural, 201
  that encompasses both the universe and the human being, 209
  transcosmic, 215
  triadic, 119
  world's. *See* world

Venice Declaration, 35, 220n48, 241
vibration, vibrations, 55–56, 81, 158
  everything is, vi, 55, 66
  frequency of, 50
  invisible, 81
  of strings, 71

War
  First World, 40
  World War II, 41, 172
world, 3–6, 9, 11, 12, 14–17, 23–30, 33, 37, 40–43, 45–46, 49, 56–58, 61–62, 66–67, 70–71, 73, 76, 78–79, 82–84, 87, 90, 97, 106, 112, 119, 121–22, 131, 133, 135, 138–39, 144, 150–52, 166–68, 171, 175, 183, 185, 191–92, 197–99, 206–7, 209, 212–15, 220n1, 223n8, 232n52, 243, 247, 249
  abstract, 58, 192
  ancient vison of the, 37
  as it is, 195
  bootstrap's, 96
  classical, 43, 185
  classical vision of the, v, 37
  coherent vision of the, 112
  complex, 11
  contemporary, 20
  contemporary vision of the, 41
  deterministic, 1
  divided, 192
  empty, 192
  exterior, 171
  feminity of the, 39
  foundation of the, 28
  four-dimensional, 78
  future, 34
  Great Text of the, 139
  hadron, 68, 91

Harmony of the, 184, 245
humanistic, 8
image of the, 34, 190
infra, 171
interior, 171
irreducible mystery of the, 215
knowable, 215
logical, 167
macrophysical, 42, 43, 58, 105, 121
macroscopic, 57
masculinization of our, 9
massacres, 40
mathematical, 192
mechanistic, 1
microphysical, 129–30
microscopic, 131
modern, 30, 142, 148, 203, 235n12, 248
monadic aspect of the, 116
multidimensional, 79
natural, 56, 71, 97
nature of the, 117
new, 27, 40
new vision of the, 27, 40
observed, 94
of appearance, 156
of art, vii, 167, 170
of art, literature and philosophy, 81
of classical science, 32
of culture, 194
of dark, 22
of events, 117, 119, 198
of irreducible interconnections, 127
of light, 22, 29
of particles, 105, 126
physics, 194
of Quantum Culture. See quantum
of quantum mechanics, 25
of science, 32
of superstrings, 70
of symbols, 32
of the marvelous, 171
of thinking, 26
old, 40
one-dimensional, 66
one possible, 224n19, 241
outer, 193, 206

outside, 87–88, 193
perfect, 62
perfection in our, 4
physical, 66, 111, 142–43
premodern, 203
presence of the sacred in the, 14, 106
present, 34
psyche, 128, 130–31, 142–43
quantum, v, vii, 27, 30, 42, 43, 45, 46, 51, 56–58, 121, 126–28, 142, 147, 151–52, 155, 171–72
quantum image of the, 150
quantum vision of the, v, 40
quantum view of the, 125–26, 133
rational, 1
rationality of the, v, 23
rational order of the, 182
real, 45, 90
scientific, 8, 174
scientific approach to the, 127
scientific vision of the, 213
sensible, 43
sensitive, 26
spontaneity of the, 13
supersensible, 198
supra, 171
systemic visions of the, 112
systemic view of the, 133
tangible, 10
temporal, 5
terrestrial, 5
theater of the, 145
three-dimensional, 81
three, 29
triadic unity of the, 119
two-dimensional, 66
unknown, 45
unification of the, vi, 58
unified view of the, 127
uniqueness of our, vi, 87
unity of the, vii, 111, 131, 183, 197
unknowable, 215
upside-down, 150
view of the, 48, 97
vision of the, 106, 122
visible, 27, 75, 200

www.ingramcontent.com/pod-product-compliance
Lightning Source LLC
Chambersburg PA
CBHW030531230426
43665CB00010B/839